CHAN

Cytochrome Oxidase
A Synthesis

Cytochrome Oxidase

A Synthesis

MÅRTEN WIKSTRÖM, KLAAS KRAB
and MATTI SARASTE

University of Helsinki, Finland

1981

ACADEMIC PRESS

A Subsidiary of Harcourt Brace Jovanovich, Publishers

London · New York · Toronto · Sydney · San Francisco

ACADEMIC PRESS INC. (LONDON) LTD
24–28 Oval Road
London NW1

United States Edition published by
ACADEMIC PRESS INC.
111 Fifth Avenue
New York, New York 10003

British Library Cataloguing in Publication Data
Wikström, M.
    Cytochrome oxidase.
    1. Cytochrome oxidase
    I. Title   II. Krab, K   III. Saraste, M.
    574.19′258   QP603.C85

    ISBN 0-12-752020-1
    LCCCN 81-67888

Typeset by Preface Ltd., Salisbury, Wilts., and printed in Great Britain by
Thomson Litho Ltd., East Kilbride, Scotland

Preface

The intention in this book is to provide a current, integrated and fairly comprehensive insight into the structure and function of cytochrome oxidase, which is, if not the most important, at least one of the most intriguing enzymes of aerobic metabolism. The present work is not a review of the voluminous literature on this subject, at least not in the usual sense. We have called our approach a synthetic one to emphasize our attempt to bring together both temporally and methodologically different research material and concepts, and to construct from this a single picture of this enzyme. Admittedly, our success in this respect is limited. It is clear that there are still considerable gaps of information, which require much future work before they can be filled. However, we have attempted to bridge many of these gaps with working hypotheses and models, which is why part of the presented material is speculative and should be viewed as such. Yet we hope that the book may prove useful also as a source of references. We apologize for the possibility that our approach may have emphasized some research data at the expense of other data of equal or even more importance. However, this is the inevitable price that must be paid in attempting a synthesis, showing, if that price is high, the authors' inadequate judgement.

With the exception of Chapters 1 and 2, which describe the scope of our work and give a bird's eye view on the subject, respectively, the book is best read in the numbered chapter sequence.

During the process of writing we have been aided by a great number of colleagues in the form of criticism, access to unpublished data, and discussions. We are particularly indebted to Drs A. Azzi, G. Babcock, C. H. Barlow, G. Buse, R. Capaldi, E. Carafoli, R. P. Casey, B. Chance, M. Clore, K. De Fonseka, L. Ernster, R. Henderson, P. Hinkle, W. J. Ingledew, D. Kell, A. A. Konstantinov, J. S. Leigh, Jr, B. Ludwig, M. W. Makinen, B. G. Malmström, B. D. Nelson, T. Ohnishi, T. Penttilä, R. O. Poyton, G. Schatz, E. Sigel, G. J. Steffens, R. J. P. Williams and E. Yang, whose invaluable help is gratefully acknowledged. In addition, several colleagues have provided us with permission to reproduce figures from their original work, which is acknowledged separately in the appropriate context.

We are also grateful for the understanding and sympathetic attitude of Arthur Bourne of Academic Press during the prolonged stages of writing the manuscript. Ms Marja Immonen, Ms Sirkka Rönnholm and Ms Hilkka Vuorenmaa have given us invaluable help in preparation of the manuscript. Our research group has been supported by grants from the Sigrid Jusélius Foundation and the Finnish Academy (Medical Research Council) during preparation of this book, K. K. was supported by a post-doctoral fellowship from EMBO, and by a travel fellowship from the Sigrid Jusélius Foundation.

Helsinki, May 1981   Mårten Wikström
                                                              Klaas Krab
                                                             Matti Saraste

It is far better to foresee even without certainty than not to foresee at all
(Henri Poincaré)

I keep the subject constantly before me and wait till the first dawnings open little by little into the full light
(Sir Isaac Newton)

Everything should be made as simple as possible but not simpler
(Albert Einstein)

On to a bridge
Suspended over a precipice
Clings an ivy vine
Body and soul together
(Bashō)

Contents

1

Scope

Modern research on cytochrome oxidase is multidisciplinary to say the least. Due to its central position in the energy metabolism of all respiratory organisms, this enzyme has attracted much interest in the fields of bioenergetics and energy metabolism. Its function as an oxidoreductase and as the principal O_2-reducing enzyme makes it an interesting subject in the fields of biological electron transport and oxygen activation mechanisms. As an integral protein in the inner mitochondrial membrane that is assembled from several polypeptide chains, cytochrome oxidase has attracted scientists in the fields of protein structure and topography. The fact that part of the cytochrome oxidase protein is coded for by the mitochondrial genome and synthesized on mitochondrial ribosomes, while part follows the more familiar nuclear-cytoplasmic route, has stimulated much research in the fields of genetics and protein biosynthesis, as well as in mitochondrial biogenesis. The presence of four different redox centres in the cytochrome oxidase molecule (two haems and two coppers) has traditionally interested haemo- and cuproprotein chemists as well as physicists due to the applicability of a variety of spectroscopic and other physical techniques to unravelling the structure of these centres. On top of this truly multidisciplinary attack, research in the cytochrome oxidase field has traditionally been divided into research on the isolated and purified enzyme in detergent solution on one hand, and on the membranous oxidase in mitochondria on the other.

From this it is clear that the cytochrome oxidase literature is not only voluminous, but that research has been and still is conducted from a great variety of angles, all of which require a certain degree of specialization. One of the greatest problems is, in our view, the almost total lack of co-ordination of the different approaches. For instance, it is not uncommon that kineticists studying the purified and solubilized enzyme are ignorant of functional characteristics observed only with the membranous enzyme. Analogously, students of mitochondrial energy conservation are often not sufficiently initiated in the kinetic and catalytic properties of the enzyme. These are but a few examples of a situation which has rather obvious causes, and which is by no means uncommon to modern experimental science in general.

Although lack of detailed knowledge from neighbouring disciplines might not be a hindrance for scientific development up to a certain level, "cross-information" is as a rule essential for further progress both conceptually and experimentally. If an interdisciplinary approach is not taken in time, there is a certain danger of stagnation in the development due to "saturation" with experimental detail.

Our feeling that this situation may be imminent in the research on cytochrome oxidase gave us the first motive to write this book. The voluminous and multifaceted literature on this enzyme has also, in our opinion, prevented researchers from drawing connections on a temporal scale. Modern intense studies, often with sophisticated new techniques, have come into the foreground, but this has sometimes happened at the expense of very useful information gathered some 10–20 years ago, or earlier.

We have called our approach in producing this book a synthetic one. By this we do not mean to claim that we have succeeded in incorporating every piece of experimental information into a singular picture, but synthesis certainly describes the general thrust and direction of our endeavour, which are encapsulated by the first three quotations on page vii. The last quotation beautifully describes our ultimate goal. It is clear, however, that this has not been achieved. Much more information will be required than is presently available to describe the structure and function of cytochrome oxidase in molecular detail so that all the information clings together in perfect harmony. But we hope that our approach might provide a stimulus for more integrated research on this enzyme in the future.

We would like to persuade the reader that the kind of approach taken is often associated with unforeseen and sometimes delightful discoveries, similar to finding a missing piece in a jigsaw puzzle. However, more important than such delights is the fact that such discoveries can and should be put to test by experiment. Experiments that have been suggested in this way are almost unique in the sense that they would not have been designed without the unifying model. It is, perhaps, mainly for this reason that we concur so wholeheartedly with Poincaré, whose statement introduces this book. The danger of "foreseeing" structure and function in terms of models and theories is, we think, compensated for by the secure settlement that can be reached by experimental test. On the other hand, the process of "foreseeing", even at the risk of failure, is not compensated for by anything. This has been our second main motivation for the approach taken in this book.

2

Introduction and general orientation

Cytochrome oxidase is the oxygen-activating enzyme of cellular respiration in eukaryotes (animal, plant and yeast cells) as well as in certain prokaryotes. In the former the enzyme is located in the inner mitochondrial membrane, and in the latter organisms it is part of the cell membrane. Cytochrome oxidase enables these cells to oxidize foodstuffs using molecular oxygen by catalysing electron transfer from cytochrome c to O_2.

The utilization of O_2 as the terminal oxidant by all higher forms of life has probably contributed greatly to evolution due to the large energetic advantages over other available oxidants. The essential nature of cytochrome oxidase may be exemplified by the fact that it is probably responsible for more than 90% of the O_2 consumption by living organisms on Earth. The very critical dependence of vital organs such as brain, heart muscle and kidney on aerobic metabolism is another facet of this enzyme's central position in physiology. As succinctly stated by Lemberg (1969) in his already classical review on cytochrome oxidase,

> "the general significance of cytochrome oxidase thus greatly exceeds that of haemoglobin, its much studied and much more completely known chemical relative. Biologically, haemoglobin is only an auxiliary of the process of cell respiration in that it carries the oxygen into the tissues via the bloodstream. This is necessary only in bulky animals, in which diffusion of oxygen from the surface or from a tracheal system is insufficient."

The history of cytochrome oxidase research covers the entire period of modern biochemical research (Table 2.1). Due to the limited space available we cannot give a full historical account here. Such an account may also be unnecessary in view of the eloquent historical reviews available (Slater et al., 1965; Keilin, 1966; Lemberg, 1969; Nicholls and Chance, 1974; Florkin, 1975). Here we limit ourselves to a brief chronological list of "classical" discoveries on which much of our present basic knowledge rests (Table 2.1).

In the following sections we will present a condensed orientation and survey of cytochrome oxidase as it is known today. This is to aid readers who may not be familiar with this enzyme and the different aspects of its study. To save space, most sections include only a minimum number of

3

Table 2.1 Chronological list of classical events in the research on cytochrome oxidase.

1884–87	McMunn reported on the four-banded spectrum of histo- or myo-haematin in several tissues.
1924	Warburg proposed that cellular oxygen consumption is catalysed by an iron-containing enzyme, *der Atmungsferment*, ferric iron being reduced by foodstuffs and reoxidized by oxygen.
1925	Keilin rediscovered McMunn's pigments and identified them as three species, the *cytochromes a, b* and *c*.
1926–33	Warburg *et al.* showed that a haem-containing enzyme is essential for cellular respiration using cyanide and CO as respiratory inhibitors. The photosensitivity of CO inhibition was used to obtain the "photochemical action spectrum" of *der Atmungsferment*. Keilin considered an oxidase separate from the cytochromes, which he suggested might be a copper enzyme.
1929	Dixon proposed the name cytochrome oxidase.
1938	Keilin and Hartree demonstrated the essential role of cytochrome *c* as electron donor to the terminal oxidase, which they called cytochrome *c* oxidase.
1939	Keilin and Hartree showed using their microspectroscope and with the aid of several inhibitors that their previous "cytochrome *a*" was, in fact, composed of two different species, only one of which (called cytochrome a_3) reacted with ligands. The remainder of the original "cytochrome *a*" retained this name.
1953	Chance *et al.* showed that the CO–ferrocytochrome a_3 is photo-dissociable with a dissociation spectrum identical to Warburg's "photochemical action spectrum". This was the final proof for the co-identity of *der Atmungsferment* and cytochrome a_3.
1954	Maley and Lardy and Lehninger showed that oxidation of cytochrome *c* by O_2 is coupled to oxidative phosphorylation.
1958–61	Okunuki *et al.*, Hatefi *et al.* and Griffiths and Wharton developed the methods for isolation and purification of cytochrome oxidase (which by now was the name for the cytochrome aa_3 entity).
1959–60	Although the presence of copper had previously been noted by several groups, Sands and Beinert provided the first proof for its functional role in cytochrome oxidase.

references. Only Section III.C is different in this regard since basic aspects of the proton pump are discussed, which will not be dealt with any further in subsequent chapters. Relevant information with complete quotations on material presented in this chapter, and indeed in the whole book, may be obtained from one or several of the following review articles or symposium volumes: Falk *et al.* (1961), King *et al.* (1965, 1979), Lemberg (1969), Malmström (1973, 1979), Nicholls and Chance (1974), Caughey *et al.* (1976), Capaldi and Briggs (1976), Wikström *et al.* (1976, 1981), Dutton *et al.* (1978), Erecińska and Wilson (1978), Azzi and Casey (1979), Wikström and Krab (1979*a*), Azzi (1980).

I. Metal centres

A. Nomenclature and chemistry

Cytochrome oxidase (ferrocytochrome c : O_2 oxidoreductase; EC 1.9.3.1), also called cytochrome c oxidase (sometimes cytochrome aa_3), contains two haem groups and two protein-bound copper ions per minimum catalytic unit, i.e. the aa_3 monomer. On extraction of the non-covalently bound haem from the protein, only haem A is found (Fig. 2.1). Typical features of haem A are the carbonyl group in position 8 and the long isoprenoid chain in position 2 of the porphyrin ring. Haem iron may be further liganded by two (fifth and sixth) co-ordination bonds in the axial direction, perpendicular to the plane of the ring. The haem is a planar disk with a side c. 8.5 Å long and c. 4.5 Å thick.

It is established that the two haems A of the monomer are *a priori* in very different environments. Thus the terms haems a and a_3 are clearly motivated. It is possible that the two haem groups are attached to different polypeptide chains. Although this has not been established unequivocally, "cytochromes a and a_3" is a very commonly used terminology. The main difference between the haems a and a_3 is that the latter is usually of high

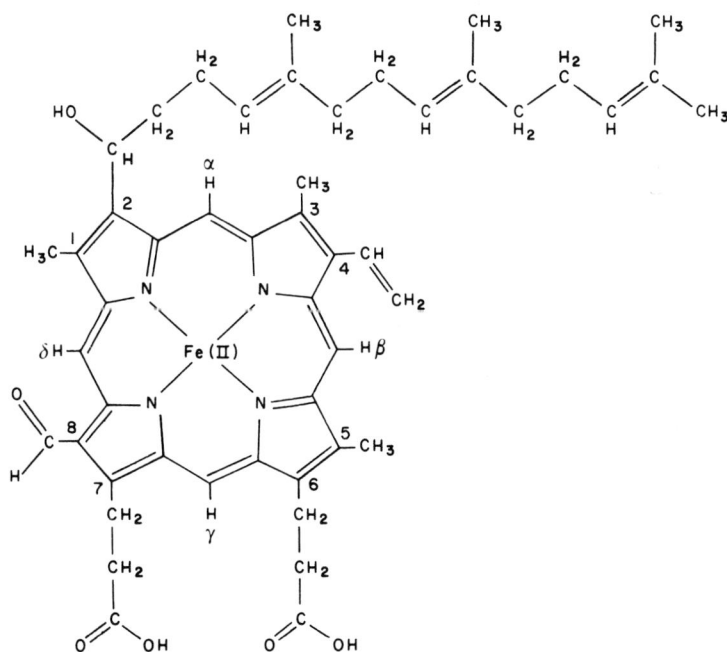

Fig. 2.1   The structure of haem A.

spin and reacts with various ligands, whereas the former is usually of low spin and does not. In fact, this is the classical definition of cytochromes a and a_3, of which the latter binds O_2, CO, etc., in the ferrous state, and HCN, HN_3, H_2S, etc., in the ferric state (Keilin and Hartree, 1939). These ligands are bound to the sixth axial position of haem a_3.

Similarly to the haems, the two copper atoms are also in very different environments. This is revealed mainly by spectroscopic and magnetochemical studies. The two copper atoms have been named in a variety of ways in the literature, e.g. Cu_{vis} and Cu_{invis} on the basis of "visibility" and "invisibility" by EPR spectroscopy, or Cu_a and Cu_{a_3} on the basis of their assumed functional and structural associations to the two haems. In this book we will use the more neutral terms Cu_A and Cu_B, of which the former is the easily EPR-detectable copper, which appears to be in rapid redox equilibrium with haem a. Cu_B is the usually EPR-indetectable copper, which is in close functional and physical contact with the haem of cytochrome a_3. The co-ordination of the two coppers is largely unknown, although some proposals have been made on the basis of EPR spectroscopy and sequence data.

B. Spectroscopy

Various spectroscopic methods have proved very useful in studies of cytochrome oxidase. The most commonly used method is optical spectrophotometry, by which oxidoreduction of the haems, in particular, may be monitored. Figure 2.2 shows the absolute optical spectra of reduced and oxidized cytochrome oxidase. In addition to the bands shown, fully oxidized oxidase exhibits a band at 820–840 nm (about 2 mM^{-1} cm^{-1} per aa_3 unit), which to at least 85% is due to Cu_A^{II} (Wharton and Tzagoloff, 1964; Boelens and Wever, 1980; Beinert *et al.*, 1980). The fully oxidized enzyme also shows a weak band at 655 nm, which has been attributed to ferric haem a_3 in its particular linkage with Cu_B.

All bands in Fig. 2.2 are attributable to haem transitions. Interpretation of these spectra in terms of cytochromes a and a_3 has been the subject of much controversy and ambiguity (see Chapter 4). However, there is presently strong evidence in favour of the original proposal (Keilin and Hartree, 1939) that the 605 nm band of the reduced enzyme is mainly due to ferrous haem a, whereas the band at 445 nm is due to both a and a_3 in roughly equal proportions.

The EPR spectrum of oxidized "resting" cytochrome oxidase, as isolated, is shown in Fig. 2.3(a). It reveals only two clearly defined components, viz. a low spin ferric haem with resonances centred at $c. g = 3, g = 2$ and $g = 1.5$, and a signal with $g = 2$, which is attributed to Cu_A^{II}. Quanti-

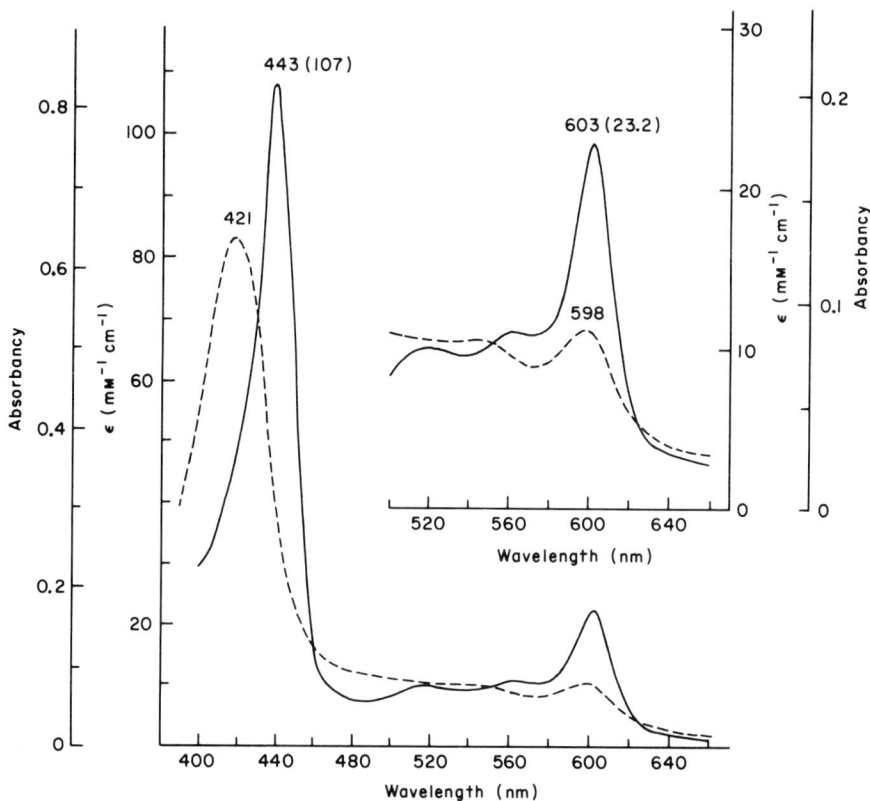

Fig. 2.2 Absolute spectra of fully reduced (——) and fully oxidized ("resting") (----) cytochrome oxidase. Extinction coefficients are on a haem A basis (should be multiplied by two to get the extinction on an aa_3 basis). From Vanneste (1966) with permission.

tation of the EPR signals reveals that the low spin haem represents only some 50% of the total haem present, and that the $g = 2$ signal due to copper accounts for only 40% of the copper that is intrinsic to the enzyme. Extraneous copper with well defined EPR characteristics is often associated to the isolated enzyme, but may be removed by dialysis against EDTA.

As shown in Fig. 2.3(b), the EPR spectrum changes dramatically on partial reduction. The low spin haem resonances disappear and are replaced by high spin ferric haem signals in the $g = 6$ region. All EPR resonances disappear on full reduction of the enzyme.

Also the EPR data have been difficult to interpret in terms of assigning

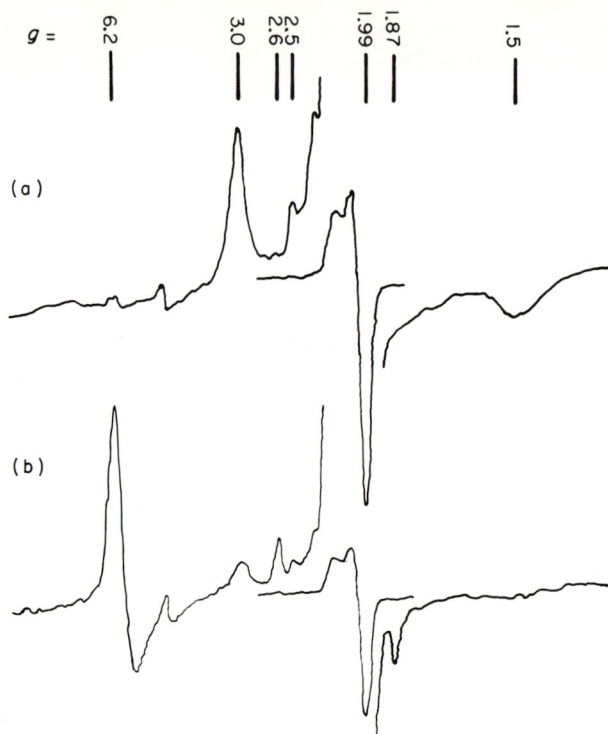

Fig. 2.3 EPR spectra of (a) fully oxidized and (b) partially reduced cytochrome oxidase. Redrawn from work by Van Gelder and Beinert (1969) with permission.

specific signals to specific redox centres. However, a thorough analysis, in which magnetic susceptibility, MCD spectroscopy and other techniques are also included, yields a reasonably consistent picture (Chapter 4).

C. Oxidoreduction properties

Anaerobic titrations of cytochrome oxidase with reducing equivalents have shown that the enzyme accepts four electrons per aa_3 unit. Potentiometric and coulometric titrations show that at least both haems and Cu_A accept one electron in the anaerobic transition from fully oxidized ("resting") enzyme to the fully reduced state. By inference, Cu_B is therefore likely to accept one electron as well; this view is supported by other, more indirect, experimental data.

Anaerobic redox titrations reveal two haem components with apparent midpoint potentials at pH 7 of about 230 and 380 mV. Identification of these haem transitions with the respective cytochromes has been another area of controversy, closely related to the difficulties of interpreting optical and EPR spectroscopy (Chapters 4 and 5). Today, however, it is almost certain that the two redox transitions cannot be equated simply with the two haems a and a_3, but rather each one represents a mixture of both haems. This is likely to be due to extensive antico-operative redox interactions.

The $E_{m,7}$ value of Cu_A is fairly well established to be near 240 mV (pH-independent), whereas the corresponding potential of Cu_B is probably much more positive, perhaps of the order of 340 mV (also pH-independent). In contrast to the E_m values of the copper ion transitions, those of the two haem transitions are both dependent on pH.

The best known and best defined states of the enzyme are the fully oxidized "resting" state and the fully reduced state, respectively. It seems likely that the haems are ferric in the former and ferrous in the latter state, and the copper ions in the Cu^{II} and Cu^I forms, respectively. Many studies have dealt with the so-called mixed valence or half-reduced state of the enzyme, which by definition contains two electrons per aa_3 unit. However, in many conditions it is difficult to define this state in terms of redox states of the four individual centres. This is due to the interpretational difficulties referred to above, but also to the lack of redox equilibration often encountered among oxidase molecules, and perhaps within aa_3 units. Only in some cases is it possible to define the mixed valence state more precisely. One such state is the much utilized half-reduced carbonmonoxy derivative. In this case it is reasonably certain that haem a_3 and Cu_B are reduced whereas haem a and Cu_A are oxidized.

II. Structure and organization in the mitochondrial membrane

Mammalian cytochrome oxidase has been reported to contain from six to 12 different subunits, but it is still uncertain how many of these polypeptides are true constituents of the enzyme. Apparently the three largest polypeptides have this status, also indicated by their being produced by mitochondrial transcription and translation. The two largest subunits (numbered I and II) may be involved primarily in "anchoring" of the redox centres, whereas subunit III may have a specific role in proton translocation. The role of the cytoplasmatically synthesized polypeptides associated to the oxidase unit as isolated (four of these are likely to be true subunits) is still obscure.

Cytochrome oxidase is a typical integral membrane protein which is "plugged through" the membrane. The natural electron donor to the oxidase, viz. cytochrome c, is located peripherally on the outer or cytoplasmic

(C) side, where it binds to specific binding sites that protrude into the aqueous phase.

The location of the redox centres is largely unknown. Studies with orientated membrane samples show, however, that the haems are orientated with their planes perpendicular to the plane of the membrane. Haem a and Cu_A may be located fairly close to the C side, whereas haem a_3 and Cu_B appear to be deeply buried and shielded by the phospholipid bilayer. It may be added that the much quoted evidence for a "transmembranous" arrangement of the redox centres is ambiguous (see Wikström and Krab, 1979a, and Chapter 7).

Lateral crystals of membranous oxidase have recently been studied by electron microscopy and image reconstruction techniques. This has given further information about the positioning in the membrane as well as about the gross shape of the enzyme particle.

III. Energetics and catalytic functions

A. Basic energetic considerations

The overall reaction catalysed by the oxidase in solution is electron transfer from cytochrome c to O_2. Cytochrome c is a pure electron donor at physiological pH (Dutton et al., 1970). Recent data interpreted to the contrary (Green and Vande Zande, 1981) were probably complicated by the redox-linked protolytic effects in cytochrome oxidase (Chapter 5; see also note 3 on p. 190). Thus the H^+ required in formation of water from reduced O_2 (i.e. one such "substrate proton" per electron) must be taken from the medium during functioning of the enzyme in solution:

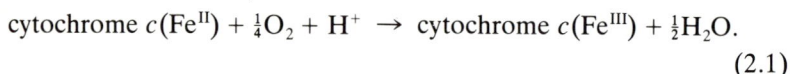

$$\text{cytochrome } c(\text{Fe}^{II}) + \tfrac{1}{4}O_2 + H^+ \rightarrow \text{cytochrome } c(\text{Fe}^{III}) + \tfrac{1}{2}H_2O.$$

$$(2.1)$$

In the membranous state it is of importance to know from which side of the membrane the "substrate" H^+ is taken up (Section III.C and Chapter 7).

At pH 7 and in air-saturated medium reaction (2.1) implies electron transfer across a redox potential span of about 500 mV. This highly exergonic reaction is thus associated with a negative free energy change of $c.$ 23 kcal (96 kJ) for each $2e^-$ transfer (or reduction of an oxygen atom).

This energy is not released as heat, however, but largely conserved for subsequent ATP synthesis. Cytochrome oxidase constitutes "site 3" of oxidative phosphorylation, at which the energy is primarily conserved as an electrochemical proton gradient across the inner mitochondrial membrane (Section III.C).

B. Basic features of electron transfer and O_2 reduction

Electrons enter the enzyme specifically via the haem of cytochrome a, but the further electron transfer sequence is not known in detail. In particular, the role of Cu_A is obscure. Yet, the sequence haem $a \rightarrow Cu_A \rightarrow Cu_B$; haem a_3 is often quoted. O_2 binds to the sixth (axial) position of iron (II) in haem a_3. However, O_2 may bind also to Cu_B^I as some recent data suggest. In any case, it seems clear that haem a_3 and Cu_B are closely associated, together forming a binuclear centre that catalyses the reduction of O_2 to water. The exact mechanism of this essential process is being revealed only slowly (Chapter 6). For electron transfer activity, seen note 2 on p. 190.

C. Generation of the electrochemical proton gradient

Redox energy is conserved in the respiratory chain by redox-linked generation of an electrochemical proton gradient ($\Delta\bar{\mu}_{H^+}$) across the inner mitochondrial membrane (Mitchell, 1961, 1966). Mitchell proposed that this occurs by a vectorial arrangement of hydrogen and electron transfer reactions in so-called redox or o/r loops. According to this redox loop model cytochrome oxidase generates $\Delta\bar{\mu}_{H^+}$ by catalysing electron transfer from cytochrome c on the C side of the membrane and H^+ transfer from the M side, to the haem a_3/Cu_B centre, where both are consumed in the reduction of dioxygen to water. Such a function is thermodynamically equivalent to translocation of one H^+ ion from the M to the C side per transferred electron, even though no H^+ ions would be released on the latter side (Wikström et al., 1981).

However, in 1977 it was discovered that cytochrome oxidase catalyses true proton translocation with uptake of $2H^+/e^-$ on the M side and release of $1H^+/e^-$ on the C side of the membrane (Wikström, 1977; Wikström and Saari, 1977). This overall function is summarized in Fig. 2.4. Such a proton pumping function cannot be explained by the redox loop model simply because the oxidase contains only formal electron carriers. Thus a fundamentally different kind of coupling mechanism was implicated, one that may generally be called a redox-linked proton pump (Wikström and Krab, 1978, 1979a,b, 1980; Wikström et al., 1981). The distinction between this type of mechanism, as compared with that of a redox loop, is strongly reflected in structural requirements of the catalytic proteins, in the coupling mechanism itself, and in the overall energetics. As opposed to the redox loop model of the oxidase (Mitchell and Moyle, 1967), the model depicted phenomenologically in Fig. 2.4 is thermodynamically equivalent to translocation of $2H^+/e^-$ across the membrane (see Wikström et al., 1981), and is therefore two times more efficient energetically than the redox loop model.

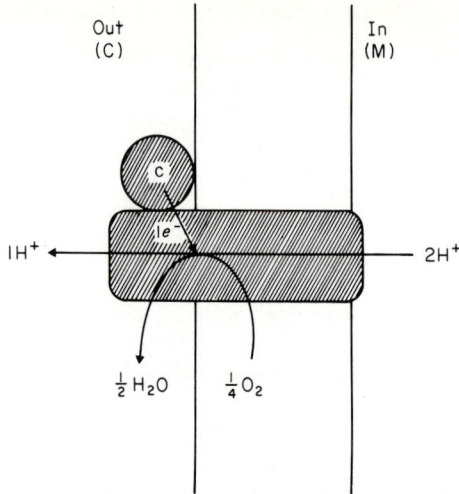

Fig. 2.4 Schematic view of proton translocation by cytochrome oxidase. The point of O_2 reduction has been drawn arbitrarily near the C side of the membrane, but may also be located closer to the M side. Uptake of H^+ from the M side and release on the C side show the net observed stoicheiometry (per transferred electron). If the "substrate" protons required to reduce dioxygen to water are taken from the C side, the proton pump must translocate $2H^+/e^-$ across the membrane in order to preserve the overall stoicheiometry observed and shown in the figure. For further details, see the text.

The proton pump notion was subsequently supported by demonstration of this function also in proteoliposomes containing the purified enzyme (Wikström and Saari, 1977; Krab and Wikström, 1978), as subsequently confirmed and extended by others (Casey *et al.*, 1979; Sigel and Carafoli, 1979, 1980; Coin and Hinkle, 1979; Sigel, 1980; Capaldi, 1981; Prochaska *et al.*, 1981).

There are three kinds of difficulties purporting to the proton pump of cytochrome oxidase that are sufficiently frequent to require some discussion.

The first difficulty is experimental and relates to the remaining controversy on the translocational H^+/e^- stoicheiometry in the respiratory chain (see Wikström and Krab, 1979a, 1980; Wikström *et al.*, 1981). On one hand, some workers still do not accept the evidence for the oxidase as a proton pump (Mitchell and Moyle, 1979; Lorusso *et al.*, 1979; Papa *et al.*, 1980), and, on the other, two groups claim a higher stoicheiometry of proton translocation than indicated in Fig. 2.4 (Alexandre and Lehninger, 1979; Azzone *et al.*, 1979).

The second difficulty is conceptual and relates to a later misunderstanding of the term "proton translocation" as used by Hinkle *et al*. (1972) and Hinkle (1973) in describing their pioneering experiments with cytochrome oxidase proteoliposomes. In most of these early experiments a hydrogen donor was used to reduce cytochrome c on the outside of the membrane whereby the observed release of $1H^+/e^-$ has a trivial explanation. The reason why true proton translocation was not observed was probably related to poor intravesicular buffering power (see Wikström and Saari, 1977; Wikström and Krab, 1979a). These early data were indeed quoted as the strongest evidence for the redox loop model of the oxidase at the time, with no indication of true proton translocation as in Fig. 2.4.

The third difficulty is thermodynamic and is exemplified, for instance, by the statement of Seiter and Angelos (1980) that "H^+ ejection into strongly buffered cellular interiors is electrochemically futile". This is true, of course, unless (as in Fig. 2.4) the released proton originates from the aqueous space on the other side of the membrane and, therefore, the proton translocation is associated with translocation of electrical charge. It must be re-emphasized that cytochrome oxidase has been demonstrated to catalyse translocation of *two* electrical charge equivalents across the membrane per transferred electron (Wikström, 1978; Sigel and Carafoli, 1978, 1979, 1980; Coin and Hinkle, 1980; Sigel, 1980; Krab and Wikström, 1979), in accordance with Fig. 2.4. Moreover, the uptake of $2H^+/e^-$ from the M side (cf. Fig. 2.4) was also demonstrated experimentally (Wikström and Saari, 1977).

Finally, it should be stressed that the actual mechanism of proton translocation is ambiguous in Fig. 2.4, which is purely a phenomenological summary of experimental observations. Two possible extreme interpretations are (i) that the proton pump catalyses translocation of $2H^+/e^-$ and, consequently, that the proton utilized in formation of water from reduced dioxygen must be taken from the C side of the membrane, or (ii) that the proton pump catalyses translocation of $1H^+/e^-$ and, consequently, the "water proton" is taken from the M side. Only very recent data may enable us to choose between these alternatives (Chapter 7). Clearly, this distinction is essential for the understanding of the molecular mechanism of the proton pump.

3

Structure and topography

I. Introduction

Cytochrome oxidase is one of the mitochondrial enzymes that are assembled by an interplay between the mitochondrial and the nuclear genes. In *Neurospora* (Sebald *et al.*, 1973) and in *Saccharomyces* (Mason and Schatz, 1973; Rubin and Tzagoloff, 1973*b*) the three largest polypeptides of cytochrome oxidase are of mitochondrial origin, while the four smaller ones are synthesized on cytoplasmic ribosomes. Experiments with cultured mammalian cells (Yatscoff *et al.*, 1977; Hare *et al.*, 1980) as well as with isolated mitochondria (Bernstein *et al.*, 1978; Rascati and Parsons, 1979*b*) have shown that the two or three largest subunits of the mammalian enzyme are also made on mitochondrial ribosomes. Three cytochrome oxidase genes in yeast (see Bonitz *et al.*, 1980) and mammalian mitochondrial DNA (Anderson *et al.*, 1981*a,b*) have recently been identified and sequenced.

The products of mitochondrial protein synthesis constitute only about 5% of the total mitochondrial protein (for reviews see Schatz and Mason, 1974; Borst and Grivell, 1978). Nevertheless, the mutants lacking active mitochondrial DNA are incapable of synthesizing a functional respiratory chain or an ATP synthase. On the other hand, nuclear mutants in which the formation of active mitochondrial enzymes is affected have also been characterized (Nargang *et al.*, 1979; Bertrand and Werner, 1979). Clearly, there is a co-operation between cytoplasmic and mitochondrial protein synthesis in the assembly of mitochondrial enzymes such as cytochrome oxidase into the inner mitochondrial membrane.

II. Purification of cytochrome oxidase

Cytochrome oxidase, which is assembled by several different polypeptides of different size and properties and derived from two sources of synthesis, is a complex object for structural studies. Considerable complexity is immediately apparent at the stages of isolation and purification. Although it is relatively easy to obtain high amounts of enzyme from a suitable source, such as beef heart, the quality of such preparations is not well defined by classical purity criteria. Electrophoretic analyses reveal that purified enzyme preparations often contain "minor" polypeptide components that co-purify with the enzyme. Moreover, the preparations are often polydisperse in the hydrodynamic sense. There are indications that the enzyme structure may become labile during the isolation procedure. In fact, it is not difficult to imagine that an amphiphilic membrane-bound enzyme that is composed of several polypeptides may tend to disintegrate when separated from its natural environment.

A. Methods of purification

The conventional purification procedure consists of three sequential steps. Mitochondria or preparations of submitochondrial particles are gradually solubilized through the action of a detergent. The solubilized respiratory enzymes are fractionated by ammonium sulphate precipitation. Finally the isolated enzyme is purified by repeated salting-out and dissolution steps so that the concentration of prosthetic groups rises to a maximum "desired" level. The last step is nowadays often performed by chromatography, e.g. by anion exchange in the presence of a non-ionic detergent (Mason et al., 1973), or by hydrophobic interaction chromatography (Weiss et al., 1971; Rosén, 1978b; Nagasawa et al., 1979). The "schools" of oxidase purification differ, for example, with respect to the detergent used (for a review see Hartzell et al., 1978). In most of the studies either the anionic detergents cholate or deoxycholate, or the non-ionic detergents of the Tween (polyoxyethylene sorbitan) and Triton (t-octylphenylpolyoxyethylene ethanol) series have been used.

Owing to their higher critical micelle concentrations (CMCs), the anionic detergents can be more easily removed from the enzyme preparations than the non-ionic ones (see Kagawa, 1972; Helenius and Simons, 1975). Removal of detergent has proved particularly important in studies in which the enzyme has been "reconstituted" into liposomal membranes, by which the full scale of its catalytic activities can be tested. This is why cholate (with a CMC of about 15 mM) has become a detergent of choice. Most other detergents, particularly Triton X-100, are easily retained in reconstituted liposomes (Briggs et al., 1975; Schlieper and De Robertis, 1977). This has an adverse effect on the observation of ion translocation phenomena, either because the membranes are rendered permeable to ions, or because the enzyme may preferably combine with the liposome surface rather than become properly incorporated in the membrane.

In the first stages of isolation a sequential solubilization of the membrane-bound enzymes from the mitochondria yields the best results in further purification, and simplifies the procedure (Jacobs et al., 1966a,b;

Kuboyama *et al.*, 1972). This means that instead of a complete solubilization of membrane proteins, it is more convenient to let the "stickiest" enzymes stay behind in the membrane-bound state while dissolving the less "sticky" ones. Such a membranous cytochrome oxidase has been purified free of other cytochromes by sequential extraction with detergent and salt (Jacobs *et al.*, 1966*a,b*; Sun *et al.*, 1968; Vanderkooi *et al.*, 1972).

The final purification is conventionally performed by repeated ammonium sulphate precipitations (Yonetani, 1960*a*, 1961; Fowler *et al.*, 1962; Kuboyama *et al.*, 1972). This is done in the presence of detergent and rather high detergent/protein ratios are sometimes used to achieve effective delipidation of the enzyme (Jacobs *et al.*, 1966*b*; Yu *et al.*, 1975). Among the more recently used chromatographic procedures, the most promising is affinity chromatography with immobilized cytochrome *c* (Ozawa *et al.*, 1975; Weiss and Sebald, 1978; Rascati and Parsons, 1979*a*). Based on this idea, Weiss and Kolb (1979) have developed a "completely" chromatographic isolation procedure for cytochrome oxidase from *Neurospora*. However, the chromatography procedure using cytochrome *c* bound to cyanogen bromide-activated Sepharose columns probably does not separate the mammalian oxidase on the basis of specific affinity (Weiss *et al.*, 1978; Azzi, 1981), and is neither very effective nor easily reproducible. This is presumably because the lysine residues that interact with the oxidase upon specific binding are simultaneously the most reactive ones (Brautigan *et al.*, 1978), and therefore reactive with the CNBr-activated Sepharose (Azzi, 1980). This problem was recently overcome by using cytochrome *c* from *Saccharomyces* linked to an activated thiol Sepharose column as the affinity ligand (Bill *et al.*, 1980), a procedure that may be very useful in isolation of the mammalian enzyme.

In Table 3.1 we have compiled the gross properties of several cytochrome oxidase preparations, mainly from beef heart. In spite of considerable variability in the isolation procedure, the preparations show much similarity. However, comparison of the haem/protein ratios is complicated by some major differences in the determination of haem A, none of which are apparent in the table. It is regrettable that the well established extinction coefficients determined for native haem A or its pyridine haemochromogen (Van Gelder, 1966; Lemberg, 1969; Caughey *et al.*, 1975; Nicholls *et al.*, 1976) are not used throughout, but that values previously shown to be in error are still utilized.

The polypeptide composition is also similar in the different preparations (Briggs *et al.*, 1975; Carroll and Racker, 1977; see also Fig. 3.1), although there is some variation in so-called contaminating polypeptides. The phospholipid content, on the other hand, varies considerably. This is due, in part, to potential differences between detergents in delipidation of the protein, and in part to varying extents of purification (e.g. the number of salting-out steps).

B. Some pitfalls of purification

It is important to realize that the traditional purification procedures of cytochrome oxidase probably include not only a desired destruction of the membrane, but some disintegration of the protein to be isolated as well. Electrophoresis in a porous polyacrylamide gel in the presence of a non-ionic detergent such as Triton X-100 may be used as a simple test of enzyme poly-/monodispersity (Ludwig et al., 1979; Penttilä et al., 1979; Rascati and Parsons, 1979a). Such a non-ionic detergent is benign in the sense that it does not easily dissociate protein/protein interactions. We may thus suppose that polypeptides which have an affinity to stick together (as the subunits should have in an enzyme) may be retained in complex form under such conditions (Helenius and Simons, 1975; Robinson and Capaldi, 1977).

Figure 3.2(a) shows such an electrophoretic separation of proteins and protein complexes in a conventional "purified" cytochrome oxidase preparation. Figure 3.2(b) shows the polypeptide composition of the different fractions obtained. It is clear that the protein complexes separated in the first dimension contain similar polypeptides in many cases, but in variable proportions. Many of these polypeptides have been assigned as true subunits of cytochrome oxidase. From this it may be concluded that a significant fraction of the enzyme may disintegrate during the conventional purification procedure, and that the final preparation may contain a significant fraction of enzyme protein in a modified form.

Note also that the electrophoretic fractionation in the presence of Triton yields one fraction (Fig. 3.2(b), curve D) which is at least aesthetically reasonably uniform. This fraction was selected as a goal in further purification and characterization of the enzyme, and has more recently been purified by preparative electrophoresis and anion-exchange chromatography (Saraste et al., 1980, 1981, and see below).

Heterogeneity or polydispersity is indeed quite typical of cytochrome oxidase preparations. It is apparent in most hydrodynamic studies (see Kuboyama et al., 1972; Wainio et al., 1973; Kornblatt et al., 1975) as minor components or boundary asymmetry in sedimentation velocity experiments. Kinetic analysis of the decline of molecular activity during purification has also led to the conclusion that the enzyme may become heterogeneous in isolated preparations (Vanneste et al., 1974). Usually the measured activity of isolated preparations, conveniently expressed as the number of reducing equivalents transferred per second per mole of cyto-

Table 3.1 Preparations of cytochrome c oxidase. Preparations A–E were obtained from bovine heart mitochondria, preparation F from rat liver, preparation G from baker's yeast and preparation H from *Neurospora*. Haem A/protein ratios cannot be compared directly, because different extinction coefficients have been used by different authors (see footnote).

Preparation		Detergents used	Phospholipid (mg mg^{-1})	Haem A (nmol mg^{-1})		Number of major bands in SDS-PAGE	References
A	1	Cholate	0.2	11.0	(a)	7	Kuboyama et al., 1972
	2	Cholate	0.07	10	(d)	7	Yu et al., 1975
B	1	Deoxycholate	n.d.	8.4–8.7	(b)	n.d.	Fowler et al., 1962
	2	Deoxycholate, cholate	0.17–0.25	9.4–10.6	(c)	7	Capaldi and Hayashi, 1972; Downer et al., 1976
	3	Deoxycholate, cholate	n.d.	9.5–10.5	(a)	7	Van Buuren, 1972; Rosén, 1978
C	1	TX-114, TX-100	0.16–0.18	8.0–8.8	(d)	6–7	Jacobs et al., 1966a; Sun et al., 1968; Briggs et al., 1975
	2	TX-114, TX-100	0.02	9.0	(d)	6–7	Jacobs et al., 1966b; Sun et al., 1968; Briggs et al., 1975
	3	TX-114, cholate	≤0.05	13–14	(a)	n.d.	Hartzell and Beinert, 1974

D	Deoxycholate + affinity chromatography	0.33	13.6	(b)	6	Ozawa et al., 1975, 1979
E	Cholate + hydrophobic chromatography	0.02	10	(a)	6–7	Rosén, 1978
F	TX-100, cholate, Tween-20	0.04	9.5 ± 0.9	(a)	6–7	Höchli and Hackenbrock, 1978
G	Cholate, TX-100 + DE-cellulose	0.02	9.5–10.3	(b)	7	Mason et al., 1973; Poyton and Schatz, 1975a
H	TX-100 + affinity chromatography	0.02	6.5–9	(a)	7	Weiss and Kolb, 1979

Haem determination: (a) $\Delta\varepsilon_{605}^{red-ox} = 12\ \mathrm{mM}^{-1}\ \mathrm{cm}^{-1}$; (b) $\Delta\varepsilon_{605-630}^{red} = 15.6\ \mathrm{mM}^{-1}\ \mathrm{cm}^{-1}$; (c) pyridine haemochromogen $\Delta\varepsilon_{587-620} = 21.7\ \mathrm{mM}^{-1}\ \mathrm{cm}^{-1}$; (d) $\Delta\varepsilon_{605}^{red-ox} = 13.5\ \mathrm{mM}^{-1}\ \mathrm{cm}^{-1}$. n.d. = not determined.

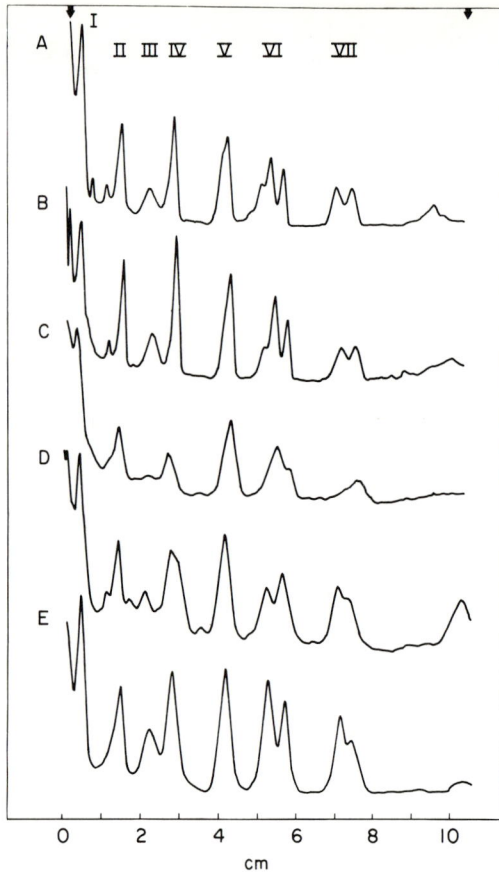

Fig. 3.1 Polypeptide compositions of different bovine cytochrome oxidase preparations. Trace A, a preparation purified according to Kuboyama *et al.* (1972); trace B, according to Capaldi and Hayashi (1972); trace C, according to Steffens and Buse (1976); trace D, according to Rosén (1978); and trace E, according to Van Buuren (1972). Electrophoreses were carried out in the presence of SDS and urea (Downer *et al.*, 1976). We are grateful to Drs R. Capaldi (preparation B), G. Buse (preparation C) and B. Karlsson (preparations D and E) for sending us samples of the enzyme purified in their laboratories.

Fig. 3.2 Two-dimensional electrophoretic analysis of a cytochrome oxidase preparation. A conventional cytochrome oxidase preparation, purified according to Kuboyama *et al.* (1972), was first electrophoresed in the presence of Triton X-100 (a). The gel was sliced as indicated, and slices A–G were analysed for polypeptide composition with electrophoresis in the presence of dodecyl sulphate and urea (b). The polypeptide pattern of the original Kuboyama preparation is shown for reference. Solid curves show the protein; the dashed curve in (a) was obtained by scanning for the presence of haem A after staining with tetramethylbenzidine. The results are taken from Penttilä *et al.* (1979).

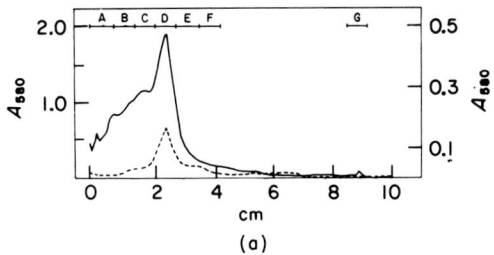

(a)

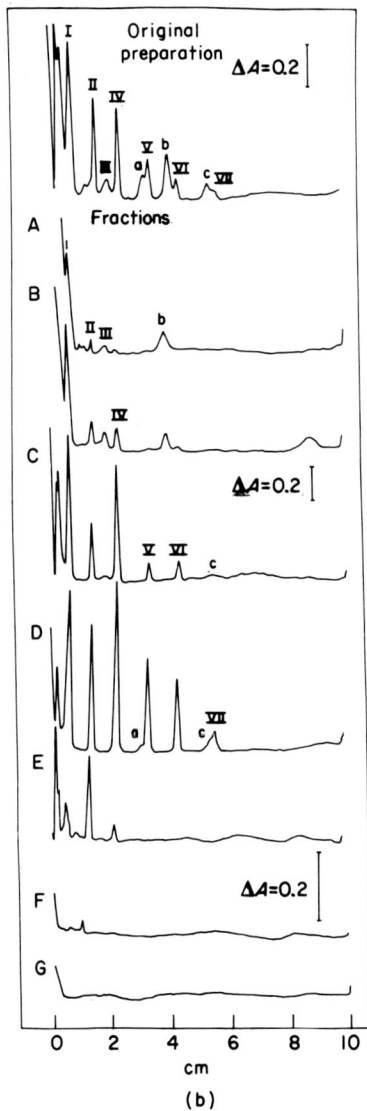

(b)

chrome aa_3 (see Hartzell et al., 1978), is lower than the activity found in isolated mitochondria. However, this difference may be more apparent than real and due, in part, to aggregation of the enzyme, inactivation by detergent, or to the difference in optimal conditions for the isolated and mitochondrial enzyme (see Nicholls and Kimelberg, 1972). Recently procedures to "reactivate" isolated cytochrome oxidase have been described (Rosevear et al., 1980).

If the enzyme is indeed not completely stable during purification, the concentration of prosthetic groups (e.g. haem A) is not necessarily a valid criterion of purity. Although there must, of course, exist a true haem A/protein ratio for the enzyme, this remains unknown until the correct molecular weight of cytochrome aa_3 has been determined accurately. The haem A/protein ratio is affected, not only by inaccuracies in haem and protein determination, but also by the polydispersity of the isolated enzyme (for collection of data see Caughey et al., 1976; Hillman and Wainio, 1977). We wish to emphasize that this ratio should not be used uncritically in the manner "the higher the number, the better the enzyme preparation". Some preparations with very high haem/protein ratios (see Komai and Capaldi, 1973) must today be regarded as purification artefacts in which the haem concentration exceeds that of the optimal purity state, and which may be subfractions of the disintegrated enzyme.

Unfortunately, partial enzyme disintegration during purification is but one source of the observed polydispersity or inhomogeneity. As discussed below in more detail, most preparations of the mammalian enzyme contain up to 12 (and possibly more) different polypeptide chains, of which several are likely to be proteins of the inner membrane that for some reason co-purify with the "true" enzyme. It is reasonable to suppose that the enzyme in situ, however complex it may be, is a clearly defined entity in the membrane with a distinct border to its environment. Preparations of "membranous" oxidase (C-1 in Table 3.1) contain some of the enzyme in a state regular enough to form lateral crystalline arrays in the membrane plane (Jacobs et al., 1966a; Vanderkooi et al., 1972). However, the polypeptide composition of "membranous" oxidase, as a whole, shows that it is less pure than preparations obtained with cholate or deoxycholate (Briggs et al., 1975; Hackenbrock, 1977; Höchli and Hackenbrock, 1978; Fuller et al., 1979). These membranous preparations are actually fragments of the inner mitochondrial membrane in which local foci of crystallization of cytochrome oxidase have formed. From the electron micrographs it is possible to say "where the enzyme ends and the membrane begins" (Poyton and Schatz, 1975a), but unfortunately this distinction becomes ambiguous when true subunits are to be distinguished among the polypeptides present in the preparation as a whole. This is indeed a major problem, at least with the mammalian enzyme, and the source of considerable controversy at present.

It may be thought that immunoprecipitation of the enzyme from solubilized mitochondria with specific antibodies would be the most straightforward tactic available to avoid the difficulties and ambiguities discussed above (see, for example, Ebner *et al.*, 1973; Mason *et al.*, 1973). However, this method cannot be used to its full potential because of an intrinsic problem, viz. the uncertain purity of the antigen used in raising the antibody. The purity of the antigen could be enhanced by using a single purified subunit in immunization. Unfortunately, however, antibodies to single subunits seem to precipitate the holoenzyme with very low recovery (Poyton and Schatz, 1975*b*). In addition, as observed by Ludwig *et al.* (1979), even when the electrophoretically purified enzyme is used as antigen, immunoprecipitation yields variable results with respect to the polypeptide composition of the precipitate. The concentration of the solubilizing detergent (Triton X-100 was used) seems to affect the result.

In conclusion, the decision on the basic question of how many subunits there are in cytochrome oxidase is beset with many non-trivial difficulties. How can we distinguish between true subunits of the enzyme and co-purified polypeptides associated merely by "artefactual togetherness", as expressed by Tanford and Reynolds (1976). One way would be to rely on analogies between oxidases from different species. The more homogeneous preparations from yeast and *Neurospora* have already been used as models of the mammalian enzyme in this respect. However, a final resolution of the latter cannot rely on such analogies. The remaining possibility is very laborious, viz. controlled dissociation and re-association of polypeptides with complete assessment of function as well as structure in the membranous state. As will be discussed below, this work has already been started with cytochrome oxidase, although it is still at its very beginning.

III. The subunits of cytochrome oxidase

Although the true subunits of the enzyme have not yet been unambiguously defined, the situation may, after all, be less of a cause for pessimism than the previous section might suggest. Some polypeptides can be assigned as true subunits with reasonable certainty. The primary structure of many of these has been elucidated. In addition, the topographical relationships are beginning to be understood. Data is also accumulating, albeit slowly, on probable functional aspects of some polypeptides (subunits).

In this section we will discuss the current data on this subject with emphasis on the mammalian enzyme. However, data from the microbial enzymes will also be mentioned, but mainly for comparative purposes, and when information on some point is lacking for the mammalian enzyme. Here we follow the "law of similarity of living creatures on the molecular level", according to which cytochrome oxidases of the aa_3 type from all organisms should exhibit functional and structural analogies. Whether such arguments are convincing or not must at present be left to the reader to decide. Only after much work in the future will it be possible to assess whether this "law" is applicable to the cytochrome oxidases to the extent that we suggest.

A. Nomenclature and number of different subunits

Identification of cytochrome oxidase subunits has had a rather tentative character. After the SDS/PAGE method for determination of polypeptide composition became available (Shapiro et al., 1967; Weber and Osborn, 1969), there was an early tendency towards simplification and a minimum number of proposed subunits (Chuang and Crane, 1971; Komai and Capaldi, 1973), although the opposite approach was also represented (Keirns et al., 1971).

In most cases the main criterion for assessment of the number of subunits has been the number of "major" bands revealed by the electrophoretic procedure. Six "major" bands were first observed in the more recent studies (see, for example, Briggs et al., 1975), but a seventh "major" band was subsequently discovered after improvement of the electrophoresis procedure (Downer et al., 1976). This newly discovered band corresponds to the third largest "subunit" (termed subunit III), which co-migrates with subunit II in conventional SDS/PAGE, aggregates easily under some conditions and combines irreversibly with subunit I under others (see, for example, Hare et al., 1980). This led Downer et al. (1976) to propose the "seven subunit hypothesis" for mammalian cytochrome oxidase, which at present is the most commonly used basis for the subunit nomenclature. However, some confusion has arisen from the fact that SDS/PAGE of purified enzyme preparations actually reveals many more bands than seven. In fact, the bands I–VII are not even the "major" ones on the basis of staining intensity under all conditions (see Figs 3.1–3.3; see also Bucher and Penniall, 1975; Steffens and Buse, 1976; Eytan and Broza, 1978). It is agreed that a total of 12 polypeptides can be discerned (Downer et al., 1976; Merle and Kadenbach, 1980; Buse et al., 1980). The current discussion is thus focused on how many of these are true constituents of cytochrome oxidase (subunits).

Downer et al. (1976) concluded that three polypeptides (called a, b and

c) with apparent M_r values of 13 000 10 400 and 8500 daltons, were "minor" components which could not be ascribed as true constituents (subunits) of the enzyme. This was based on variability in their appearance, and more recently on their removal by proteolysis without effect on the enzyme's spectral or catalytic properties (Ludwig et al., 1979; see also below). This leaves $12 - 3 = 9$ polypeptides. However, while Capaldi's group has been quite aware of the presence in "subunit VII" of three different polypeptides (band VIII of Steffens and Buse, 1976), they have preferred the simplifying "seven subunit hypothesis", perhaps mainly by analogy to the yeast enzyme. Schatz and his co-workers (Mason et al., 1973; Poyton and Schatz, 1975a,b) have presented strong evidence for the yeast enzyme being composed of seven different subunits. These form a homogeneous physical entity which is retained in chromatography and centrifugation, and which is precipitated with an antibody to a single polypeptide.

In contrast, Merle and Kadenbach (1980; see also Buse et al., 1980) in particular stress that all 12 polypeptides are true subunits of the enzyme on the basis that they are present in enzymes from different sources. Clearly, we are now facing one of the main problems discussed in Section II. It seems to us, however, that a further assessment of whether a polypeptide is a true constituent or not should not be exclusively based on which polypeptides co-purify during isolation. What is required is data on dissociation of polypeptides combined with critical checks of retention or loss of functional or structural properties of the enzyme. On the other hand, it also seems clear that we cannot rely entirely on a presumed analogy between the yeast and mammalian enzymes. After all, as recently shown by Ludwig and Schatz (1980; and see below), cytochrome aa_3 from Paracoccus contains only two different subunits after isolation, but retains several essential properties of the apparently much more complicated mammalian or yeast enzymes.

Thus the proteolytic removal of three polypeptides (see above) must be given considerable weight. Unfortunately, this test was not performed by Merle and Kadenbach (1980). Although it seems likely that their 12 polypeptides are the same as the 12 observed by others, it is difficult to be certain because polypeptides of the enzyme behave quite differently in different applications of SDS/PAGE (Capaldi et al., 1977; Tracy and Chan, 1979), and there is the possibility of differences in the enzyme preparations used (but see Fig. 3.2). However, some recent data might help to resolve this dilemma.

In our own studies, in which we have striven for isolation of a more homogeneous enzyme species, we managed to dissociate several polypeptides from the original enzyme and to clean other inhomogeneities from

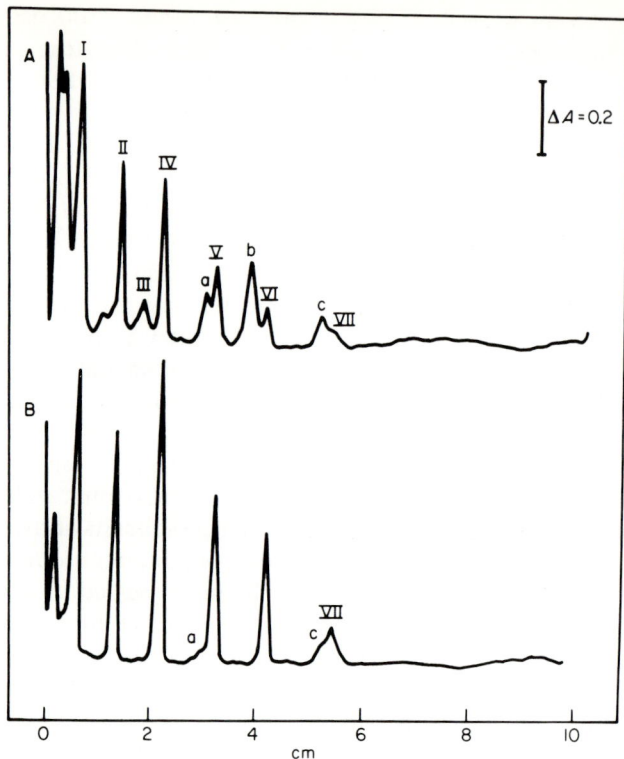

Fig. 3.3 Comparison of the polypeptide composition of a conventional and a "purified" cytochrome oxidase preparation. Trace A, cytochrome oxidase purified according to Kuboyama *et al.* (1972); trace B, this preparation further purified by preparative electrophoresis in the presence of Triton X-100 (Saraste *et al.*, 1980*a*). The polypeptides were analysed using polyacrylamide gel electrophoresis in the presence of SDS and urea (see Downer *et al.*, 1976).

the preparation (see Figs 3.1 and 3.3; Penttilä *et al.*, 1979; Saraste *et al.*, 1980). The resulting homogenous (see Section IV) enzyme preparation lacks not only the "minor" components a, b and most of c, but also the component III, which is a likely subunit (see below). Although we have called this a "six subunit" preparation, band VII is still heterogeneous. The name is, however, in keeping with the seven subunit nomenclature. This nomenclature seems preferable at the present time (whether the "seven subunit hypothesis" is correct or not) to avoid further confusion due to differences in nomenclature (see also Azzi, 1980).

The "six subunit enzyme" retains the major spectroscopic and electron transfer properties of the parent species. A preparation with similar polypeptide composition was also described by Carroll and Racker (1977) after chymotryptic digestion. In both cases the enzyme can, moreover, be successfully incorporated in liposomal membranes with retention of excellent respiratory control. The only function that is apparently lacking is net proton translocation (see Section III.D and Chapter 7).

Taken together, all this data suggests that polypeptides a, b and c, as well as "subunit III", are rather easily removed by different procedures, and were therefore the first to be suspected not to be true constituents of the enzyme (Penttilä et al., 1979). Professor Kadenbach and his collaborators kindly performed an SDS/PAGE study using their highly resolving procedure (Merle and Kadenbach, 1980) on the "six subunit" preparation. This is shown in Fig. 3.4, and may be most helpful in the achievement of identification of the "minor" components a, b and c with the polypeptides of Merle and Kadenbach, as called for above.

Figure 3.4 confirms that the "six subunit" preparation has lost subunit III. It also shows that two of the three bands running in position VI are lost, and further that one of the four bands running in positions VII–VIII has been removed. In work by others, only three bands have been discerned in the latter region (see review by Azzi, 1980, and above). In addition, band V is split, which we have also observed under some conditions (but see Fig. 3.3, trace B). Comparison of Fig. 3.4 with trace B in Fig. 3.3 suggests that the "minor" components a, b and c are identical with bands VIa, VIb and VIIa, respectively, in the nomenclature of Merle and Kadenbach (1980). Hence, the only difference to the results from Capaldi's group discussed above seesm to be the presence of an additional polypeptide in the lowest molecular weight region (called VIII by Merle and Kadenbach).

Since the role of components a, b and c is thus at least uncertain (see above seems to be the presence of an additional polypeptide in the lowest enzyme are probably to be found among the polypeptides denoted I, II, III, IV, V, VI and VII, of which VII, however, contains at least two but probably three different polypeptides. Although it might turn out in later work that not all these nine polypeptides are true constituents of the enzyme, this conclusion is clearly in support of the "seven subunit hypothesis", after the simplification concerning the status of VII.

Table 3.2 summarizes data on the seven "subunits" of the bovine heart enzyme. Their individual chemical and topographical properties are discussed below. Polypeptide heterogeneity of isolated oxidase preparations has been discussed in detail by Steffens and Buse (1976), by Boonman (1979) and by Azzi (1980).

B. The stoicheiometry of the subunits

Any structural model of the enzyme must rely on the relative subunit stoicheiometry. Many investigators have calculated the relative molar

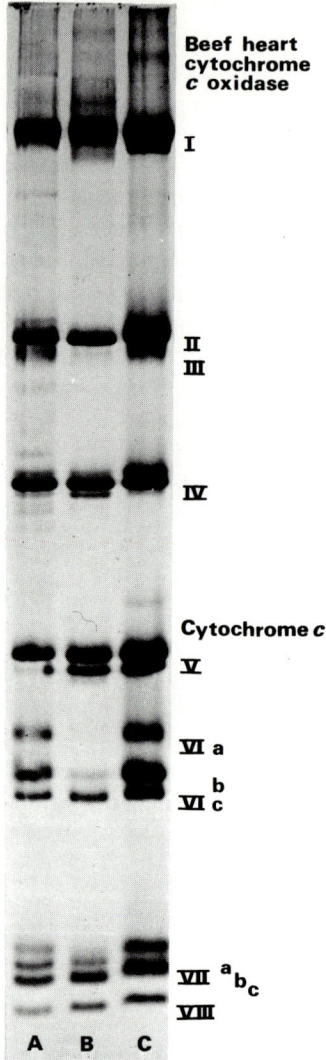

Fig. 3.4 Comparison between three cytochrome oxidase preparations by SDS/PAGE. A and C are preparations isolated in the laboratories of Kadenbach and Buse (see text for references). Preparation B is the "six subunit enzyme" prepared by Saraste *et al*. Courtesy of Professor B. Kadenbach.

Table 3.2 Subunits of the cytochrome oxidase.

Nomenclature	Approximate molecular weight in SDS/urea-PAGE ($\times 10^{-3}$)	Comparative molecular weight determinations[†] ($\times 10^{-3}$)		N-terminal amino acids[‡]	Remarks[§]
		(a)	(b)		
I	35.3 ± 2.0	35.0	56.5‖	f-Met-Phe-Ile-Asn	Hydrophobic
II	25.2 ± 0.4	23.0	26.3	f-Met-Ala-Tyr-Pro-Met	Amphiphilic, sequenced
III	21.0 ± 0.8		28.7‖	(Met)-Thr-His-Gln-Thr-His	Hydrophobic
IV	16.2 ± 0.8	17.0	17.2	Ala-His-Gly-Ser-Val	Amphiphilic, sequenced
V	12.1 ± 0.8	12.5	12.4	Ser-His-Gly-Ser-His	Hydrophilic, sequenced
VI	6.7 ± 0.9	9.7		Ala-Thr-Ala-Leu	Hydrophilic
VII	3.4 ± 0.2	5.3	5.4	Ser-His-Tyr-Glu-Glu	Amphiphilic, sequenced¶

[†] Determinations listed in column (a) have been performed by gel chromatography in SDS or guanidinium hydrochloride (Briggs et al., 1975; Downer et al., 1976). In column (b) the values have been calculated from the sequence data for subunit II (Steffens and Buse, 1979), subunit IV (Sacher et al., 1979), subunit V (Tanaka et al., 1979), and subunit VII (Buse and Steffens, 1978).

[‡] The N-termini (Steffens and Buse, 1976) were identified with the seven subunit protein with the help of Drs G. Buse and G. Steffens.

[§] "Amphiphilic" means that the protein contains a long, continuous hydrophobic sequence and/or is not water-soluble in isolated form.

‖ Molecular weights calculated from the amino acid sequences predicted by the structure of mitochondrial DNA (Anderson et al., 1981b).

¶ Band VII contains three different polypeptides in apparently equimolar amounts (Buse and Steffens, 1978). Data presented here refer to VII_{ser}.

stoicheiometries from scanned electrophoresis patterns after staining for protein (e.g. Keirns *et al.*, 1971; Briggs *et al.*, 1975; Yu and Yu, 1977). However, protein staining is usually not suitable for this purpose because the staining intensity is quantitatively related to protein mass only for identical or very similar proteins. Each protein–stain complex has a unique molar absorptivity, the accurate determination of which is difficult (see Marres and Slater, 1977).

The fact that the probably correct (see below) one-to-one stoicheiometry hypothesis has been adopted from such data must be regarded as coincidental. In fact, it is not supported by the distribution of staining intensity in a homogeneous enzyme preparation (see Penttilä *et al.*, 1979). Quantitative N-terminal amino acid analyses mentioned by Steffens and Buse (1976) also indicated a one-to-one stoicheiometry between subunits. However, an independent determination of the stoicheiometry in the mammalian enzyme appeared warranted.

Saraste *et al.* (1980) used radioactive methyl acetimidate to label all lysyl residues and free N-termini of the enzyme. Incorporated radioactivity was subsequently determined from the separated polypeptides, both in the original enzyme preparation and in the "six subunit" preparation (see above). This data has been compiled in Table 3.3. The one-to-one stoicheiometry between subunits is particularly obvious for the "six subunit" enzyme. Only band VII yields a result significantly different from this, which no doubt is due to its heterogeneity (see above). Also the polypeptide stoicheiometry of the original enzyme preparation conforms to this picture, although the variations are greater. Only subunit III appears to be present in one-half of the molar amount as compared to the others. This may have a trivial explanation, such as loss of a significant portion of subunit III during isolation. On the other hand, it could mean, if confirmed, that there is only one copy of subunit III per two sets of the others (cf. Chapter 7).

Sebald and co-workers (1973) studied the molar stoicheiometry of the three mitochondrial subunits of the *Neurospora* enzyme. From determination of the amino acid composition and incorporation of radioactive leucine, they concluded that these subunits (i.e. I, II and III) were present in a 1 : 1 molar ratio. These studies were later extended to resolve the stoicheiometry of all seven subunits, and the presented data closely fit a 1 : 1 ratio (Weiss and Sebald, 1978).

There is thus strong evidence both with the mammalian and with the microbial enzymes that seven putative subunits are present in a one-to-one molar ratio. No doubt, this strengthens the postulated role of the corresponding polypeptides as true subunits of the enzyme. However, as discussed above, some uncertainty remains with respect to subunit III and the

Table 3.3 Stoicheiometry of the subunits in bovine cytochrome oxidase. Denatured subunits of cytochrome oxidase were modified by reaction with ^{14}C-labelled methyl acetimidate. The reaction resulted in total modification of the ϵ-amino groups in lysines and of the free N-termini. The amount of incorporated radioactivity was measured after electrophoretic separation of the polypeptide in SDS/urea-PAGE. Stoicheiometry is calculated relative to subunit II. The number of expected reactive groups per polypeptide has been calculated from the published sequences of subunits I, II, III, IV, V and VII (see Table 3.2), and from the amino acid composition for subunit VI (see Steffens and Buse, 1976; Steffens et al., 1980). Results from Saraste et al. (1980).

Subunit	Expected number of reactive sites	A. Preparation purified according to Kuboyama et al. (1972)† (see Fig. 3.3, trace A)		B. Preparation purified with TX-PAGE† (see Fig. 3.3, trace B)	
		CPM/site	Stoicheiometry	CPM/site	Stoicheiometry
I	9	319	0.98	214	1.01
II	6	327	1.00	212	1.00
III	3‡	175	0.54	—§	
IV	19	229	0.70	221	1.04
V	8	332	1.02	221	1.04
VI	7	383	1.00	256	1.21
VII¶	5	398	1.22	295	1.40

† Approximately 108 and 62 μg protein were analysed in A and B, respectively.
‡ Number of reactive sites is revised to 3 (see Anderson et al., 1981b) from the previous estimate (Saraste et al., 1980).
§ Subunit III protein is not present in this preparation, see Fig. 3.3, trace B.
¶ This band contains more than one polypeptide component.

polypeptides of band VII. Finally, it should be noted that the molar ratio of unity between the subunits contains no information as to how many sets of subunits might be required to constitute the active enzyme (see Section IV).

C. The chemistry of the subunits in different organisms

Most (and perhaps all) of the polypeptides present in preparations of purified cytochrome oxidase have been isolated and characterized in denatured form (Downer *et al.*, 1976; Steffens and Buse, 1976). The three heaviest polypeptides corresponding to subunits I, II and III are of mitochondrial origin, of which I and III are found to have a particularly hydrophobic character.

An evolutionary link between the mitochondrial cytochrome oxidases has been clarified by determination of mitochondrial DNA sequences. Cytochrome oxidase subunits I, II and III in the yeast, bovine and human enzymes exhibit very highly conserved homologies (Coruzzi and Tzagoloff, 1979; Fox, 1979; Thalenfeld and Tzagoloff, 1980; Bonitz *et al.*, 1980; Anderson *et al.*, 1981*a,b*). For the case of the hydrophobic subunits I and III, not only the charged (acidic or basic) residues are conserved, but there are also almost identical hydrophobic stretches in the sequences of corresponding subunits from different organisms. This evolutionary constraint indicates that both these aspects are important in the three-dimensional folding of these subunits, and in their assembly into the enzyme complex.

There are, however, also differences between the mitochondrial subunits from different organisms. Comparison of the primary structure of the mammalian subunit II (Steffens and Buse, 1979; Barrell *et al.*, 1979) with the corresponding gene sequence in yeast (Coruzzi and Tzagoloff, 1979; Fox, 1979), and with the primary structure of this subunit in *Neurospora* (Machleidt and Werner, 1979), shows that the two latter proteins are synthesized in a precursor form (cf. Severino and Poyton, 1980). The precursor has an *N*-terminal extension, 15 amino acid residues long, that is apparently cleaved off post-translationally.

The subunit II sequence shows some interesting features that may give a clue to its function (Section III.D). It may be noted here that the recent genetic work on mitochondrial DNA has, together with the sequence work on subunit II, revealed interesting aberrations from the genetic code, previously thought to be universal. Thus UGA is, for instance, not a stop codon but codes for tryptophan in mitochondrial DNA.

The DNA nucleotide sequence of the subunit III genome was recently elucidated (Thalenfeld and Tzagoloff, 1980; Anderson *et al.*, 1981*a,b*). The deduced amino acid sequence contains seven stretches of almost exclusively hydrophobic residues, which, if they traverse the membrane as α-helices (see Azzi, 1980; Wikström et al., 1981), would make this protein intriguingly similar to bacteriorhodopsin. There may be an even more intriguing similarity of this subunit to the DCCD-binding protein in the F_0 segment of the H^+-translocating ATPase. Thus subunit III also binds DCCD (Casey et al., 1980; Steffens et al., 1980; Prochaska et al., 1981) to an isolated amino acid carboxyl group (glutamic acid) in the middle of one of the strongly hydrophobic sequences, as shown for the DCCD-binding protein (Sebald et al., 1980; see review by Wikström et al., 1981). These similarities may indeed by directly related to similarities in function (see Section III.D and Chapter 7).

The amino acid sequence of subunit IV (147 residues) is mainly hydrophilic, but includes a segment of 19 hydrophobic amino acid residues (Sacher et al., 1979). This cytoplasmically synthesized subunit is almost certainly located on the matrix side of the inner mitochondrial membrane (Section V), which is interesting considering its site of synthesis.

Subunit V, which was sequenced by Tanaka et al. (1979), is a water-soluble protein with 109 amino acid residues. The sequence was reported to resemble the β-chain of haemoglobin, but the extent of overlap is hardly sufficient to support this proposal.

Some of the low molecular weight polypeptides in cytochrome oxidase preparations have been sequenced by Buse and his co-workers (Buse and Steffens, 1978; Buse et al., 1980; Steffens et al., 1979). One of the sequenced polypeptides in band VI (see Steffens et al (1979), who termed this VII; this is VII_{ala} in Azzi's (1980) nomenclature), having a molecular weight of about 10 000 daltons, is probably identical with the "minor" component b (Fig. 3.2), and is thus probably extrinsic to the enzyme (see Buse et al., 1980).

The N-termini of three polypeptides of band VII have been determined (see Buse et al., 1980; this corresponds to their band VIII), and these have consequently been termed VII_{ser}, VII_{ile} and VII_{phe}, respectively (Azzi, 1980). Of these, VII_{ser} was sequenced by Buse and Steffens (1978). It contains 47 residues with one continuous hydrophobic segment of 20 residues.

Table 3.4 compares the subunit compositions of cytochrome oxidases of different origin, as determined under roughly similar experimental conditions. The comparison demonstrates that "corresponding" subunits from different enzymes are not electrophoretically identical although the general subunit patterns are similar (cf. Poyton et al., 1978). The subunit structures of the purified bacterial cytochrome oxidases from Thermus (Fee et al., 1980) and Paracoccus (Ludwig and Schatz, 1980) appear much simpler than those of the mitochondrial enzymes (Table 3.4). The possible

Table 3.4 Subunit composition of cytochrome c oxidase purified from different organisms. The subunit compositions as analysed with SDS/PAGE (in the absence of urea). Results for *Thermus thermophilus* HB8 (Fee *et al.*, 1980), *Paracoccus denitrificans* (Ludwig and Schatz, 1980), *Saccharomyces cerevisiae* (Rubin and Tzagoloff, 1978), *Neurospora crassa* (Weiss and Kolb, 1979), rat liver and bovine heart (Höchli and Hackenbrock, 1978) enzyme are listed.

Subunit	Apparent molecular weight ($\times 10^{-3}$)					
	Thermus	*Paracoccus*	*Neurospora*	*Saccharomyces*	Rat[†]	Bovine[†]
I	55.0	45.0	40.0	40.0	43.5	41.5
II	33.0	28.0	29.0	27.3	24.5	22.5
III	—	—	21.0	25.0	—	—
IV	—	—	18.0	13.8	15.5	15.0
V	—	—	14.0	13.0	12.0	13.2
VI	—	—	12.0	10.2	10.5	10.2
VII	—	—	9.0	9.5	9.0	7.5

[†]The subunit III protein, when present, co-migrates with band II in SDS/PAGE. Note that the molecular weight values differ from those presented in Table 3.2.

significance of this with respect to enzyme structure and function is discussed below (Section III.D).

Figure 3.5 shows an electrophoretic comparison of bovine heart, yeast and bacterial cytochrome aa_3 complexes (Ludwig, 1980). It also includes data on immunological cross-reactivity between polypeptides, which may be of interest if analogies between different enzymes are to be expected. It is noteworthy that not only are the subunit II antibodies cross-reacting with the putatively analogous subunit of the other species, but some of the small molecular weight polypeptides of cytoplasmic origin seem to cross-react with antisubunit II from *Paracoccus* and *Saccharomyces*. Hence there is an apparent common origin of subunits II, but a genetic relationship between mitochondrial and cytoplasmic subunits is also possible.

D. Binding of prosthetic groups and other functions of subunits

A number of efforts have been made to locate the redox centres in cytochrome oxidase to specific subunits. This is, however, very difficult because denaturation and separation of subunits as a rule causes dislocation of the redox centres. The results obtained after staining the electrophoresis gels for the presence of copper or haem (Gutteridge *et al.*, 1977) have shown that many of the bands have the ligands attached to them. Distribution of haem A among the partially separated subunits and subunit complexes in gel permeation chromatography (Phan and Mahler, 1976*b*; Yu and Yu,

**Cross-reaction with
anti-subunit Ⅱ
(*Paracoccus*)**

**Cross-reaction with
anti-subunit Ⅱ
(*Saccaromyces*)**

s　　b　　p　　　　s　　b　　p　　　s　　b　　p

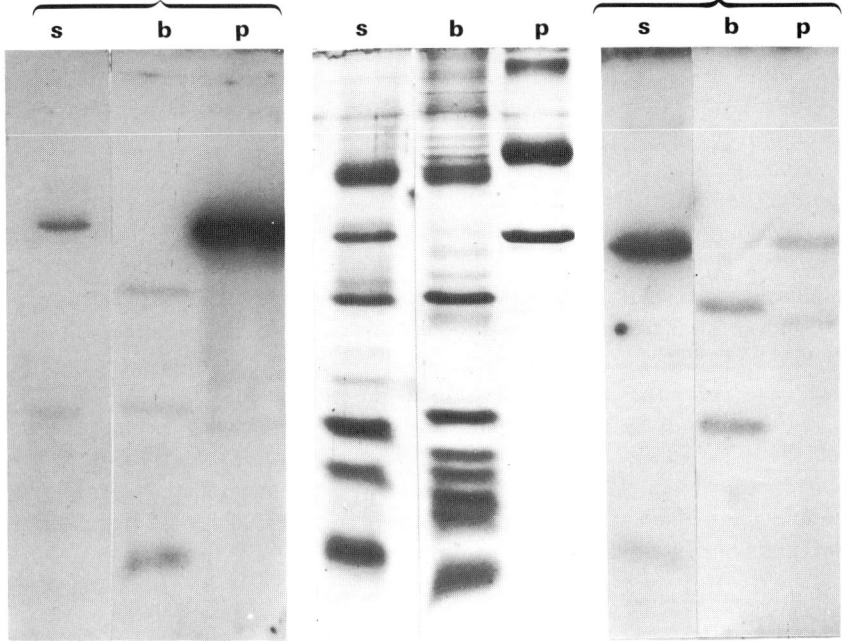

Fig. 3.5 Comparison of subunits of the different cytochrome oxidases. The subunits of *Saccharomyces* (s), *Bovis* (b) and *Paracoccus* (p) cytochrome oxidases were resolved by SDS/PAGE in 15% gel (middle). Immunological cross-reactivities of the separated subunits with the *Paracoccus* (left) and *Saccharomyces* (right) subunit II antisera were determined as described by Ludwig (1980); positive cross-reactivity is detected using autoradiography. Results taken from Ludwig (1981) courtesy of Dr B. Ludwig.

1977), or isoelectric focusing (Freedman *et al.*, 1979) is liable to very similar difficulties of interpretation. Haem A, partially or fully dislocated from its native environment, may complex with any suitable binding site found in the SDS–protein solution. Co-migration of haem and copper with one or several subunits after destruction of the enzyme cannot be extrapolated to solve the localization problem (but see below for some more recent possible success).

However, isolation of a haem-containing subunit (or a copper-containing one) can be valuable when specificity in the complexation can be indicated.

Yu *et al.* (1977) purified a haem–protein complex from the denatured enzyme by extraction with 50% pyridine—a procedure that may not fulfil the requirements called for above. The purified haemoprotein turned out to contain subunit V and was highly enriched in haem A (40 nmol mg^{-1} of protein). This complex was apparently composed of 2 mol of the subunit per mole of haem. Could this mean that the haem might be sandwiched between two copies of the subunit V protein as if the latter contained one-half of a haem-binding site?

Analogously, MacLennan and Tzagoloff (1965) isolated a copper-binding subunit from succinylated cytochrome oxidase. Purification resulted in a single cuproprotein with an approximate M_r of 25 000 daltons (subunit II?), containing 1 mole of Cu per 12 300 daltons. Could this suggest that subunit II might bind two coppers (cf. below)?

One approach to gaining insight into possible haem or copper binding would be to search, using structural and phylogenetic arguments, for amino acid sequences that may be candidates for such functions.

Subunit II has an homologous amino acid sequence with copper proteins of the blue (Type I) plastocyanine family (Steffens and Buse, 1979; Yasunobu *et al.*, 1979). The homologous sequence includes sulphuric amino acids which are probably required in binding of copper. This particular sequence in subunit II is as follows:

Human   -Tyr-Tyr-Gly-Gln-Cys-Ser-Glu-Ile-Cys-Gly-Ala-Asn-His-Ser-Phe-Met-...

Bovine   -Tyr-Tyr-Gly-Gln-Cys-Ser-Glu-Ile-Cys-Gly-Ser-Asn-His-Ser-Phe-Met-...

Yeast   -Phe-Tyr-Gly-Ala-Cys-Ser-Glu-Leu-Cys-Gly-Thr-Gly-His-Ala-Asn-Met-...

The substitutions found in the human (Barrell *et al.*, 1979) and yeast (Coruzzi and Tzagoloff, 1979; Fox, 1979) proteins are conservative and do not affect the proposed copper-binding ligands (underlined), viz. two cysteines, one histidine and one methionine (Buse *et al.*, 1978). In plastocyanine, which has an homologous sequence, but lacks a second Cys in the vicinity of the first one, it is known that a copper ion is bound to the apoprotein via two histidines, one methionine and one cysteine (see Colman *et al.*, 1978).

Subunit II is clearly a good candidate for a copper-binding site in cyto-chrome oxidase. Whether it binds one copper or two (as the data of MacLennan and Tzagoloff, 1965, might suggest) is not known. Recent data by Winter *et al.* (1980; see below) indicate that all copper in the enzyme might be bound to subunit II, but since their experiments were not performed quantitatively, any firm conclusion on this point would be premature.

Although no conclusive evidence is yet available, several recent findings suggest that subunits I and II may be primarily involved in binding of all four redox centres. The isolated *Paracoccus* enzyme contains only two different subunits and appears to be identical or closely similar to the mitochondrial enzyme with respect to electron transfer activity and spectroscopic parameters (see above and Albracht *et al.*, 1980). Moreover, the properties of the two subunits resemble those of subunits I and II of the mitochondrial enzyme (Ludwig, 1980; Section II.C). Secondly, subunits I, II and III are all of mitochondrial origin. However, subunit III is not required in electron transfer, and its removal has little effect on the optical spectrum (Penttilä *et al.*, 1979; see below and Chapter 7). Thus if we may assume that the subunits of mitochondrial origin are most essential for catalytic activity, these results again indirectly point at subunits I and II as candidates for binding the redox centres. Thirdly, Winter *et al.* (1980), who recently employed a controlled detergent procedure for subunit dissociation, found that haem A was exclusively associated to subunits I and II, and that all detected copper (unfortunately not quantified; see above) was associated with subunit II.

If this line of thought is followed, the topography of subunits I and II (Section V.B), as compared with what is known about the topography of haems a and a_3 (Chapter 4), would lead to the hypothesis that haem a_3 might be bound to subunit I and haem a to subunit II. This contention is further supported by combination of the facts that subunit II provides the binding site for cytochrome c (Section V.C) and that haem a is the immediate acceptor of electrons from haem a (Chapter 6).

Although this hypothesis appears quite suggestive, it should still be borne in mind that each redox centre may be bound to two or more polypeptide chains in the native enzyme. Two or more subunits could well be involved in providing the necessary binding domains of the haems in particular, which are expected to exhibit a number of interactions with the protein "pocket", including van der Waals' contacts and hydrogen bonds. The structural difference between the mitochondrial and bacterial cytochrome oxidases (Table 3.3) might hence only be apparent. Here the "law of similarity, etc." (Section III) could be operating at the level of tertiary (and, indeed, of quaternary) structure. If, to take an *ad hoc* example, one haem A were bound exclusively by the comparatively large subunit II in the *Paracoccus* enzyme, the analogous binding domain in the mitochondrial enzymes could be made up in part by subunit II, and in part by one or more of the subunits of cytoplasmic origin. This could furnish an explanation for the observed immunological cross-reactivity between subunit II antibody and cytoplasmic subunits (Section III.C), and for the enigmatic function of at least some of the latter. Although we do not wish to stress this hypothesis too strongly due to its highly speculative character, we may note that subunit V was shown to bind haem A under artificial conditions. In line with our speculation, this cytoplasmic subunit could thus participate in formation of a haem-binding domain, together with either subunit I or II.

There are now several indications for the hypothesis that subunit III may be primarily involved in proton translocation by cytochrome oxidase (Chapter 7). This is of great interest for several reasons, not least because this may be the first subunit of the enzyme that has been dissociated without denaturation of the remaining enzyme (Penttilä *et al.*, 1979; Saraste *et al.*, 1980; cf. above). Possible functions of this subunit are discussed in more detail in Chapter 7, also with respect to the fact that it is apparently missing from the isolated bacterial enzymes (Table 3.4).

We have now compiled suggestions for possible functions of subunits I, II, III and V. Of the remaining subunits, IV and VII_{ser} both have a single hydrophobic stretch of about 20 amino acids. Buse and Steffens (1978) suggested that polypeptides of this type might have a structural role, perhaps by "anchoring" the enzyme to the membrane and/or by provision of connections between subunits. To this we may add the possibility of connections to other catalytic proteins in the membrane, although the presence of such connections does not at present have much experimental support.

IV. Quaternary structure of cytochrome oxidase

The assembly of the subunits into the cytochrome oxidase complex must be studied with the isolated enzyme, which requires use of detergents as an essential tool. Under suitable conditions the state of an isolated membrane protein in detergent solution can be assumed to be reminiscent of the native membranous state (see Tanford and Reynolds, 1976). But changes in catalytic activity sometimes observed during isolation (Vanneste *et al.*, 1974) should serve as a warning that significant structural alterations might also take place.

A. Minimal molecular weight: the haem aa_3 unit mass

One of the crucial factors that determines the molecular size of cytochrome oxidase is the protein mass corresponding to one haem A. When this is known, the minimal molecular weight, which corresponds to the size of the two-haem monomer (aa_3), can be determined. The required parameter is clearly the ratio between haem (or haem iron) or intrinsic copper (see Chapter 4), and protein (cf. Section II).

The determination of these parameters is not only beset with analytical difficulties, but both the presence of impurities (i.e. extrinsic co-purified proteins) and lability of the enzyme during purification (see Section II) have made interpretation of the data most difficult.

The sum of molecular weights of the seven subunits in a one-to-one stoicheiometry is about 130 000 daltons for the bovine heart enzyme when M_r values are used that have been determined by SDS/PAGE and amino acid sequencing (when available). The size of subunits I and III, as predicted from the corresponding gene structures (Anderson et al., 1981b; cf. Table 3.2), seems to be remarkably larger, and would increase this sum up to about 160 000 daltons. These values correspond to haem/protein ratios around 6–8 and 13–15 nmol mg^{-1} of protein, respectively, depending on whether the aa_3 unit is composed of two or one sets of the subunits. When this prediction is compared with the data compiled in Table 3.5, it is immediately seen that the observed haem/protein ratios often fall between the predicted values, particularly for the mammalian enzyme. Although values as high as 14 nmol of haem per milligram of protein have been reported occasionally (see Hartzell and Beinert, 1974), the value is usually found to be near 10.

This apparent discrepancy led us to search for alternative answers. One possible explanation could be that subunits are present in different molar amounts (Wikström et al., 1978). However, the recent hard data on the subunit stoicheiometry (Section III.B) makes this highly unlikely. In our next approach we had realized the heterogeneity of most enzyme preparations and consequently strove for isolation of a homogeneous preparation to be able to make a more critical assessment of the enzyme's gross quaternary structure. After much exploratory work, we found that the enzyme could be cleaned up considerably by gel electrophoresis in Triton X-100 (Penttilä et al., 1979; see Figs 3.1 and 3.3). However, the apparently homogeneous preparation still had a haem/protein ratio near 10 nmol mg^{-1}, contained only six "major" subunits (see Section III), and had a particle size of about 210 000 daltons (see Section IV.B). These findings inevitably led us to propose a "protomer" model, according to which the aa_3 unit would be composed of two sets of the "major" subunits (Saraste et al., 1980), with some interesting structural consequences. Fortunately, the error was detected soon afterwards, and is an educating one. The initial "six subunit" preparation was found to contain considerable amounts of apoprotein, that could be separated from the haem-containing holoprotein by gel permeation chromatography (Saraste et al., 1981). This explains the "clean" appearance of the preparation in SDS/PAGE (Fig. 3.1).

After due separation of the apoprotein, the haem/protein ratio of the purified "six subunit" preparation rose to about 19 nmol mg^{-1} (Table 3.5),

Table 3.5 Minimal molecular weight of cytochrome oxidase.

Enzyme	Haem A/protein[†] (nmol mg^{-1})	Apparent molecular weight per haem A ($\times 10^{-3}$)	Molecular weight from the subunit composition[‡] ($\times 10^{-3}$)	References
Rat liver	9.49 ± 0.91	105 ± 11	115 (6)–127 (7)	Höchli and Hackenbrock, 1978
Bovine heart	9.70	103	107 (6)	Briggs et al., 1975
	9.6–11.3	89–104	122	Downer et al., 1976
	19 ± 2	53 ± 6	–127 (7)	Saraste et al., 1981
Saccharomyces	9.5–10.3	97–105	107 (6)	Poyton and Schatz, 1975a
	14.4–15.0	67–69	142 (7)	Rubin and Tzagoloff, 1973a
Neurospora	14–15	67–71	139 (7)	Sebald et al., 1973
	6.5–9	111–154	142 (7)	Weiss and Kolb, 1979
			144 (7)	
Paracoccus	26.4–29.1	34–38	73 (2)	Ludwig and Schatz, 1980
Thermus	10.0 ± 1.3	100 ± 15	88 (2)	Fee et al., 1980

[†] The haem A determinations have been performed with different procedures (see references).

[‡] The minimal molecular weight was calculated from the sum of the apparent molecular weights of the six or seven subunits (the reported values have been obtained with SDS/PAGE or SDS/urea-PAGE) assuming one-to-one stoicheiometry. The number of the different subunits is indicated in parentheses.

which would correspond to a molecular weight of 105 000 daltons for the aa_3 unit. This is close to 106 000, which is the sum of the apparent molecular weights of the individual subunits I, II, IV, V, VI and VII (Table 3.2). It is thus clear that the aa_3 unit comprises one of each subunit.

It is no longer difficult to explain why cytochrome oxidase preparations usually exhibit haem/protein ratios in the vicinity of 10 nmol mg^{-1}. We recall that the conventional preparations actually contain 12 different polypeptides (Section III.A), many of which are probably extrinsic to the enzyme. In addition, some of the haem A may be deleted during purification, and protein complexes may dissociate and reassemble in odd proportions (Fig. 3.2). The non-uniformity is not only expressed by the haem/protein ratio but, in particular, by hydrodynamic measurements (Saraste et al., 1981). The conventional bovine heart cytochrome oxidase preparations are probably mixtures of various complexes that differ in their individual polypeptide compositions and/or stoicheiometries.

Note also that in the cases of Saccharomyces and Neurospora there is no agreement on the minimum weight (Table 3.5). The results obtained by Weiss and Kolb (1979) indicate that there would be one haem for each set of seven different subunits, while Sebald et al. (1973) proposed that the haem/subunit ratio is 2 : 1 in the purified Neurospora enzyme. This difference is analogous to the difference between the now retracted "protomer" model and our present conclusion on the mammalian enzyme (see above). The published data on the yeast enzyme (Poyton and Schatz, 1975a; Rubin and Tzagoloff, 1973a) suggest a haem/subunit ratio between 1.5 and 2, and those on the Paracoccus enzyme a ratio of 2 (Ludwig and Schatz, 1980). On the other hand, the results with the enzyme from Thermus thermophilus (Fee et al., 1980) suggest that the corresponding ratio may be close to unity.

In view of all the difficulties involved and the now quite unambiguous results with the mammalian enzyme, it seems likely to us that there is one set of subunits per aa_3 (two-haem monomer) in all oxidases of this type.

B. The size of the functional unit

The minimal catalytic unit, cytochrome aa_3, has classically been considered also to be the functional unit of the enzyme in the membrane (see, for example, Lemberg, 1969). Earlier hydrodynamic studies of solubilized cytochrome oxidase preparations are difficult to assess because detergent binding to the enzyme was usually not determined. When a hydrophobic protein to which detergent or lipid is associated is studied, e.g. by sedimentation analysis, the effect of the bound molecules should be accounted for if a significant error is to be avoided (Tanford et al., 1974).

Robinson and Capaldi (1977) determined the molecular weight of the bovine enzyme in both deoxycholate and Triton X-100 solutions. The enzyme was carefully prechromatographed on a gel column to measure detergent binding. Molecular weights were calculated to be 345 000 daltons in Triton and 200 000 daltons in deoxycholate after correction for detergent binding. The authors estimated the minimum M_r per haem to be 90 000 daltons and concluded consequently that the enzyme is mainly dimeric in Triton and monomeric in deoxycholate.

Sedimentation analyses performed in a number of laboratories have generally indicated that the molecular size of the monomeric complex is close to 200 000 daltons, but there is unfortunately rather large scatter in the results (see review by Capaldi and Briggs, 1976), in part due to neglect of detergent binding. Another source of uncertainty, and of scatter of the data, is the variation in the amount of "minor" polypeptides (see Section III.A and below) associated with the enzyme complex.

We have recently approached this problem anew, taking advantage of the high degree of homogeneity of the "six subunit" preparation (see above). The particle size of this species in Triton X-100 is 210 000 daltons after correction for bound detergent (Saraste et al., 1981). This value fits remarkably well with the molecular weights determined from the subunit composition and the haem content (see Table 3.5), and shows unambiguously that the "six subunit" enzyme is dimeric in Triton X-100. This molecular size was also supported by parallel analyses with gel permeation chromatography (Saraste et al., 1981). It should be recalled that this preparation lacks the subunit III protein (see above), bringing the molecular size of a "seven subunit" dimer to about 260 000 daltons. This is still much less than the molecular size of the 350 000 dalton particle as previously determined (Robinson and Capaldi, 1977). However, the latter value has been confirmed for the dimeric preparation which still contains the "minor" polypeptides (Saraste et al., 1981). The latter polypeptides clearly affect the mean size of the protein particles in these preparations.

Comparative analysis (Phan and Mahler, 1976a) has shown that Saccharomyces cytochrome oxidase is slightly larger than the bovine enzyme. Sedimentation equilibrium studies with the enzyme from Neurospora have yielded a molecular weight of c. 300 000 daltons in Triton X-100.

From the data in this and the preceding sections it now seems possible to reach some basic conclusions on the structure of cytochrome oxidase. The monomeric two-haem unit (i.e. cytochrome aa_3) most probably consists of a single set of subunits. The present data seem to exclude proposals of subunits in different proportions (see Wikström et al., 1978), or of a "protomeric" one-haem unit with one set of the subunits (Penttilä et al., 1979; Saraste et al., 1980). It may be stressed that up until now the available data did not allow an unambiguous exclusion of such models.

The model reached here for the mammalian enzyme (and see Ludwig *et al.*, 1979) has also been proposed for the enzymes from *Paracoccus* (Ludwig and Schatz, 1980), *Saccharomyces* (Rubin and Tzagoloff, 1973*a*) and *Neurospora* (Sebald *et al.*, 1973). However, more recent data with *Neurospora* cytochrome oxidase (Weiss and Kolb, 1979) and with cytochrome oxidase from *Thermus* (Fee *et al.*, 1980) would be more in line with a "protomer" model (see above and Saraste *et al.*, 1980). More research seems to be required in these instances to sort out the apparent discrepancies.

To our knowledge there is no unambiguous demonstration of a monomeric cytochrome oxidase preparation. Love *et al.* (1970) reported that cytochrome oxidase with an estimated M_r of 200 000 daltons was split, apparently into 100 000 dalton units, by treatment with non-ionic detergent and alkali. A similar dissociation takes place after extensive succinylation of lysyl residues (Hillman and Wainio, 1977). However, in these experiments it is not certain that the dissociation product is an homogeneous species. The apparent monomerization might actually be a splitting into a mixture of subfragments. The recent data obtained by Wilson *et al.* (1980) on cytochrome oxidase from sharks are unfortunately also beset with this ambiguity. The sedimentation velocity experiments were performed at a very high enzyme concentration (for discussion, see Tanford *et al.*, 1974) and without correction for detergent (Tween 80). In our opinion these data do not allow the conclusion that true monomeric species of the enzyme were studied.

As shown by Robinson and Capaldi (1977) and confirmed in this laboratory (M. Saraste, unpublished), the "monomeric" cytochrome oxidase in deoxycholate is certaintly not monodisperse. True monomeric units have probably been met unambiguously so far only in a certain crystalline state (Fuller *et al.*, 1979; see below).

The dimeric enzyme is not only encountered in non-ionic detergent solutions, but is also observed in lateral "membrane crystals" (Henderson *et al.*, 1977). It remains possible that the dimer is the functional form of the enzyme, and that attempts at dissociating it lead to further splitting rather than to stable monomers. Some functional data also suggest that cytochrome oxidase may be dimeric in its native state (Chapter 7).

C. Resolution of cytochrome oxidase into subdomains

Resolution of the enzyme into hydrophobic and hydrophilic domains has been attempted by controlled denaturation in dilute SDS (Phan and Mahler, 1976*b*; Yu and Yu, 1977; Osawa *et al.*, 1979). However, this approach probably leads to a slow denaturation of a continuous or dynamic character, producing unstable subunit complexes that are further dissociated into free polypeptides (M. Saraste and T. Penttilä, unpublished). The same difficulty is probably present in fractionation of the oxidase into hydrophilic and hydrophobic pools of subunits by extraction with organic solvents (Fry *et al.*, 1978) or organic acids (Fry, 1979). There is no proof that any subdomain of the native enzyme is retained intact in such procedures, or that the observed partition is not simply the consequence of different solubility properties of individual denatured polypeptides.

Hydrophilic and hydrophobic, or membrane-excluded and membrane-embedded, domains (see Section V) need not be exlusively confined to individual subunits. The entire enzyme complex traverses the membrane completely (Hackenbrock and Miller-Hammon, 1975; Henderson *et al.*, 1977; Blaisie *et al.*, 1978). This suggests the presence of three subdomains, viz. inner and outer hydrophilic plus membranous hydrophobic. However, amphiphilic subunits may well be involved in formation of separate domains simultaneously; some may even be distributed between all three subdomains (Ludwig *et al.*, 1979; Capaldi, 1981).

V. Structural arrangement of cytochrome oxidase in the membrane

Information on the structure of the membrane-bound enzyme comes from various lines of investigation. Structural data have, for instance, been obtained from physical studies on crystalline cytochrome oxidase membranes (Vanderkooi, 1974; Henderson *et al.*, 1977); surface labelling of the isolated or membrane-bound enzyme has provided information about subunit topology (Eytan *et al.*, 1975; Briggs and Capaldi, 1977; Cerletti and Schatz, 1979; see also Azzi, 1980); immunochemical methods have been used to study the exposition of subunits on the two membrane surfaces (Chan and Tracy, 1978; Frey *et al.*, 1978). We will attempt a summary of this knowledge here, and try to present a comprehensive model.

A. Subunit relationships in the isolated enzyme

A simple approach to resolving the arrangement of individual subunits with respect to one another in a multi-subunit protein is chemical cross-linking. Cleavable cross-linking agents have made this type of "nearest neighbour" analysis possible. Cross-linked products and unreacted subunits are separated on SDS/PAGE, followed by, for example, reductive cleavage and subsequent analysis of previously cross-linked units by electrophoresis in the second dimension. The technique affords a qualitative study, or mapping, of neighbour relationships among the subunits. However, there are also limitations pertaining to reagent specificity and pertur-

Table 3.6 Nearest neighbour relationship of the bovine cytochrome oxidase subunits. Results of cross-linking experiments with dithio-bis-succinimidyl propionate (DSP), dimethyl-3,3'-dithio-bis-propionimidate (DTBP) and N,N'-bis-(succinimidyloxy-carbonylpropyl)tartrate (DSPT) are taken from Briggs and Capaldi (1977, 1978). Cross-linking is as indicated in the diagram, where $\sim$ indicates a cross-link:

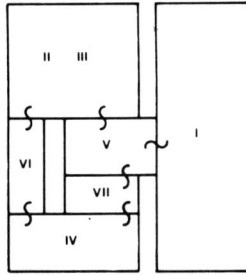

Cross-linking agent	Distance of the reactive groups (Å)	Pairs formed
DSP	11	I + V, II + V, III + V, IV + VI, V + VII, IV + VII
DTBP	11	III + V, IV + VI, V + VII
DSPT	18	II + VI

bation of the native structure making conclusions somewhat ambiguous if they rely solely on cross-linking data. Moreover, the heterogeneities in most cytochrome oxidase preparations (cf. above) as well as the possible presence of co-purified extraneous proteins certainly add to this ambiguity.

The main results of the cross-linking studies of Briggs and Capaldi (1977, 1978) are summarized in Table 3.6 and visualized by a two-dimensional map. The map is made on the basis of the raw data without consideration of possible confusion between "minor" polypeptides and "true subunits". The former are concentrated in the subunit V–VII region of SDS/PAGE (Section III), making the relationships between these particular polypeptides more uncertain. In addition, with the isolated enzyme it is possible that bound detergent can affect the reactivity of hydrophobic (parts of) subunits.

Phenomenologically, a cross-linkage may be established when two reactive (lysyl) groups are so arranged that the reagent is able to reach them both simultaneously. The data indicate that subunit V may have a central position in the enzyme in this sense (see Table 3.6). Another interesting detail from these results is the lack of cross-links between two identical polypeptides. The occurrence of such effects would strongly suggest dimerization, but its absence does not of course exclude a dimeric structure.

B. Subunit topography in the membranous enzyme

Considerable efforts have been made to find out the location of the different subunits with respect to the three phases *in situ*, viz. the aqueous phases C and M and the hydrophobic membrane phase. Reactivity of aminoacyl side chains in hydrophilic domains has been screened by water-soluble reagents such as DABS (diazonium benzene sulphonate) and NAP-taurine (*N*-(4-azido-2-nitrophenyl)-2-aminoethylsulphonate), and with antibodies to purified (denatured) subunits. Reactivity in the hydrophobic domains has been screened using lipid-soluble reagents such as azido derivatives of phospholipids.

Considering the number of difficulties and ambiguities involved in these techniques, not least with respect to preparing submitochondrial particles

Table 3.7 Topography of the subunits in the membrane-bound cytochrome *c* oxidase. Hydrophobic domains are shown as hatched areas in the diagram. The experimental data have been gathered from both microbial and mammalian enzymes. See text for references.

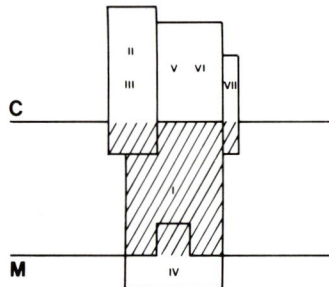

	Chemical labelling			Immunological labelling	
Subunit	Inside surface	Membrane surface	Outside surface	Inside surface	Outside surface
I	−	+	−	−	−
II	−	+	+	−	+/−
III	−	+	+	−	+/−
IV	+	+	−	+	−
V	+/−	−	+	+	+
VI	−	−	+	−	−
VII	+/−	+	+	+	+

that would be ideally "inverted" with respect to the parent mitochondria, the results are in good agreement with one another, with only a few exceptions (Table 3.7).

Eytan *et al*. (1975) and Eytan and Broza (1978) showed using DABS that the labelling pattern was very different depending on whether the enzyme was labelled from the C side or the M side of the membrane. Their data (see also Eytan and Schatz, 1975) also indicated that both subunits I and II were protected from labelling by the membrane (subunit II only partially). Later studies with DABS (Ludwig *et al*., 1979) and with antibodies (Chan and Tracy, 1978) have confirmed the apparent inaccessibility of subunit I to hydrophilic reagents (but see Prochaska *et al*., 1980), but have indicated that subunit II must be exposed on the aqueous C side of the membrane. The latter conclusion is strongly supported by the finding that subunit II provides the binding site for cytochrome *c* (Section V.C). Neither subunit I nor subunit II appears to be exposed on the M side (but see Capaldi, 1981). That both these polypeptides must be wholly, or partially, deeply buried in the membrane proper is strongly indicated by their heavy labelling by hydrophobic probes (Cerletti and Schatz, 1979; Bisson *et al*., 1979; Prochaska *et al*., 1980). Strong "hydrophobic labelling" was also found for subunit III and the polypeptides of the subunit VII group, and to a lesser extent for subunit IV, suggesting that their hydrophobic amino acid segments (see Section III) may be in contact with the membrane's phospholipid. It is interesting that neither subunit V nor subunit VI bound any "hydrophobic" label while some of the so-called impurities under bands b and c (see Fig. 3.3, trace A) were reactive.

While subunit III is certainly in considerable contact with the lipid phase of the membrane, it is also exposed to the aqueous C side (Chan and Tracy, 1978; Birchmeier *et al*., 1976; Ludwig *et al*., 1979). Ludwig *et al*. (1979) also claimed labelling of subunit III from the M side, making this polypeptide a candidate for a transmembranous structure (see also below).

Subunit IV is unambiguously labelled from the M side but not from the C side of the membrane. This is interesting in view of its synthesis in the cytoplasm. This polypeptide contains a hydrophobic sequence 19 aminoacyl groups long, the rest of the protein being highly hydrophilic (Sacher *et al*., 1979; Section III). This may be related to the labelling by hydrophobic probes, which took place equally with "deep" and "shallow" labels, differing in their extent of probing into the membrane's interior (Bisson *et al*., 1979). Subunit IV therefore might be "anchored" deeply in the membrane, although its main mass is in the aqueous M phase.

The status of subunits V–VII is much more ambiguous, which may again be related to problems of identification, as outlined in previous sections. According to the DABS-labelling data of Ludwig *et al*. (1979), subunits V and VII are labelled exclusively from the M side. According to the immunological labelling data of Chan and Tracy (1978), these are labelled from both sides, while earlier DABS-labelling experiments indicated reactivity from the C side alone (Eytan et al., 1975; Eytan and Broza, 1978). Although it is conceivable that subunit VII_{ser} may span the membrane in view of its hydrophobic stretch of 20 amino acids (Section III.D), a transmembranous position of subunit V would be difficult to reconcile with its primary structure (Tanaka et al., 1979). These conclusions are also in line with the heavy labelling of VII by the hydrophobic probes, while V remained unlabelled.

The topographical model developed on the basis of electron microscopy (Henderson et al., 1977; Frey et al., 1978; Fuller et al., 1979; Deatherage et al., 1980) has shown that about one-half of the enzyme's protein mass is located in the aqueous C phase, while only a small proteinaceous domain protrudes into the aqueous M phase (see also Fig. 3.6). The scheme in Table 3.7 is a conservative attempt to combine this information with that from the labelling data discussed above. It is quite possible that this picture may have to be revised to include transmembranous positioning of subunits II, III and VII (see Capaldi, 1981). Subunit II has two hydrophobic segments long enough to traverse the membrane in α-helical conformation (Steffens and Buse, 1979), subunit III has seven such segments (Thalenfeld and Tzagoloff, 1980; Anderson et al., 1981a,b), and VII_{ser} has one (Buse and Steffens, 1978). The data are not yet strong enough, however, to make this conclusion unambiguous. More work is needed in this area, and new results may be expected to be of considerable functional relevance (see Chapter 7).

C. The binding of cytochrome c

Cytochrome c binds non-covalently (mainly electrostatically) to the oxidase from the C side of the membrane (Jacobs and Sanadi, 1960; see review by DePierre and Ernster, 1977). From independent work in three laboratories it has been established that cytochrome c binds with a region near its exposed haem edge (see Dickerson and Timkovich, 1976) that is rich in lysyl moieties (Smith et al., 1977; Ferguson-Miller et al., 1978b; Rieder and Bosshard, 1978).

Two binding sites have been implicated per aa_3 monomer with approximate dissociation constants of 10^{-8} and 10^{-6} M, respectively (Ferguson-Miller et al., 1976, 1978b). Cytochrome c may be cross-linked to the high affinity site using dithio-bis-succinimidylpropionate (Briggs and Capaldi, 1978), or a photosensitive 3-nitrophenylazido label at Lys-13 (Bisson et al., 1977, 1978, 1980). The low affinity site was labelled by arylazido-derivatives at Lys-13 and Lys-22 (Bisson *et al.*, 1980). Erecińska (1977) also modified lysyl groups in preparing a photoreactive cytochrome *c* derivative. Birchmeyer *et al.* (1976) modified the yeast iso-1-cytochrome *c* with 5,5'-dithio-bis-(2-nitrobenzoate) at Cys-107.

Electrophoretic analysis of cross-linked products revealed that the high affinity site is on subunit II (Briggs and Capaldi, 1978; Bisson *et al.*, 1977, 1978, 1980). Subunit III may also be near the binding domain, at least in the yeast enzyme (Birchmeyer *et al.*, 1976). On the other hand, Erecińska (1977) found that small subunits (9000 and 11 000 daltons, respectively; possibly VI and V) cross-linked with cytochrome *c*. In these experiments one of the cross-linked products contained both these polypeptides plus cytochrome *c*.

Although the high affinity site for cytochrome *c* seems to be located on subunit II, other subunits may be near the binding region (e.g. subunit III), so that they may become cross-linked to *c* when a reagent is used possessing a long reactive "arm". The latter explanation may well apply to the results of Erecińska (1977). In any case, it is clear that the cytochrome *c* binding data fit well to the picture in Table 3.7.

The low affinity binding site for cytochrome *c* may be identified as lipid (presumably cardiolipin) very tightly bound to the enzyme (Bisson *et al.*, 1980). Both low affinity and high affinity sites appear to be present at a stoicheiometry of one site per aa_3 unit.

It is of considerable interest that subunit II can be implicated in high affinity binding of cytochrome *c*. Since haem *a* is unambiguously the primary acceptor of electrons from cytochrome *c* (see Chapter 6), the hypothesis is strengthened that subunit II may indeed be primarily involved in binding of this haem (see Section III.D).

D. Three-dimensional structure

Crystalline cytochrome oxidase has been prepared in two ways. A membranous preparation isolated from mitochondria using Triton (Jacobs *et al.*, 1966*a*) occasionally exhibits small crystalline areas (Vanderkooi *et al.*, 1972; Wakabayashi *et al.*, 1972; Vail and Riley, 1974). Vail and Riley (1974) observed that this occurred only when the vesicular membranes were stacked one upon another. This foreshadowed the demonstration by Henderson *et al.* (1977) that crystallization requires the interaction of enzyme molecules in the two membranes of a collapsed vesicle. In contrast, the crystalline form obtained by deoxycholate extraction of mitochondria (DOC crystals) is a monolayer (Seki *et al.*, 1970; Fuller *et al.*, 1979). DOC and Triton (TX) crystals belong to symmetry groups *pg* and *pgg*, respectively.

Figure 3.6 shows a balsa-wood model constructed by Fuller *et al.* (1979) based on electron microscopy and image reconstruction of the DOC crystals. The enzyme unit here is of about half the size of the unit in TX crystals and was interpreted to be the cytochrome aa_3 monomer. The shape is described as a lopsided Y with a length of about 110 Å, the separated domains being about 55 Å long and separated by about 40 Å. Henderson *et al.* (1977) and Fuller *et al.* (1979) suggested that the enzyme is dimeric in the TX crystal form, being composed of two units of the kind shown in Fig. 3.6 related by a twofold crystallographic axis directed perpendicular to the plane of the membrane. This interpretation was based on the assumption that the monomer contains one copy of each of seven

Fig. 3.6 Topography of protein mass of the membrane-bound cytochrome oxidase. The figure shows a model of cytochrome oxidase, produced by electron microscopic analysis of a crystal form obtained with deoxycholate extraction of mitochondria (Fuller *et al.*, 1979). M_1 and M_2 are the matrix side domains and C the cytoplasmic domain. In the other crystalline form of the enzyme produced by Triton extraction, two of these molecules are related by a twofold axis (Henderson *et al.*, 1977). The two domains M_1 and M_2 are thought to be largely embedded in the lipid bilayer. From Fuller *et al.* (1979), courtesy of Dr R. Henderson.

subunits, which indeed seems to be the best model at present (Sections III and IV).

Frey *et al.* (1978) showed that the enzyme domain that is exposed in the vesicular TX preparation reacts with anti-subunit IV but not with anti-subunit II. This important contribution has allowed an assignment of the protein into M-side and C-side domains (see Fig. 3.6). The enzyme orientation in the TX vesicles is thus "inside out" as compared to the orientation in intact mitochondria (cf. Table 3.7).

About one-half of the enzyme's mass is in the aqueous phases outside the membrane (Capaldi, 1973; Vail and Riley, 1974). However, this positioning is highly asymmetric, as revealed by electron microscopy as well as by X-ray diffraction data on oriented membranous multilayers (Blasie *et al.*, 1978). In the latter case there was no crystalline structure in the plane of the membrane. The enzyme protruded some 60 Å into the extravesicular phase. In contrast, in the TX vesicles (Henderson *et al.*, 1977), this protrusion was into the intravesicular phase. This is the cytoplasmic (C) side of the molecule (Frey *et al.*, 1978). The divided M-domain (Fig. 3.6) may stick out from the membrane only by some 15 Å. Obviously, the enzyme has the opposite orientation in the vesicles of Blasie *et al.* and of Henderson *et al.*, being in the mitochondrial configuration in the former case.

Cytochrome oxidase is clearly a transmembranous protein which is inserted asymmetrically through the membrane. The asymmetry is also seen as a large difference in the distribution of individual subunits on the two aqueous interphases (Section V.B). Surely this asymmetry must be of functional significance and probably related to the catalysis of vectorial proton translocation and charge separation across the membrane (Chapter 7). The tendency of the enzyme to form dimers under many conditions is clear. The question thus arises as to whether the dimer is the natural state of cytochrome oxidase *in situ*, as Henderson *et al.* (1977) have proposed. This intriguing problem is approached anew in later chapters, particularly in Chapter 7.

VI. Interaction with phospholipids and enzyme mobility

A. Effect of phospholipid on enzymic activity

It is well known that the oxidase is largely inactive after isolation and purification unless phospholipid or suitable detergents are added (cf. Section II). Lipid and/or detergent is, however, mostly only loosely bound to the protein, only 10 or less cardiolipin molecules being bound more tightly. Upon their removal with Triton much activity is lost, but may be restored by incubation with this phospholipid (Robinson *et al.*, 1980). Vik and

Capaldi (1977) proposed that a fluid hydrocarbon milieu is necessary for maximum electron transfer activity.

It was suggested on the basis of spin label EPR data that an "annulus" of phospholipid molecules may surround the enzyme, being relatively immobilized by the contact to the protein and being of specific importance for enzyme activity (Jost et al., 1973; Warren et al., 1975; Longmuir et al., 1977; Marsh et al., 1978). However, this suggestion was criticized by Seelig and Seelig (1978), who showed by deuterium and phosphorus NMR that cytochrome oxidase induced more disorder rather than immobilization of phospholipid fatty acid moieties upon incorporation of the enzyme into membranes. In view of this a reinterpretation of the EPR data might prove necessary (see, for example, Chapman et al., 1979).

At present it seems to us that there is no evidence for a specific lipid requirement for cytochrome oxidase activity, apart from the more general requirement of a suitable "solvent" (in this case two-dimensional) with general physicochemical properties such as to allow any mobility that may be required and to prevent aggregation of the protein, which otherwise occurs easily. A possible exception to this may be the role of tightly bound cardiolipin in providing a low affinity site of cytochrome c binding (see Section V.C and Chapter 6).

B. Mobility of cytochrome oxidase in the membrane

Hackenbrock and his collaborators (Hackenbrock, 1977; Höchli and Hackenbrock, 1978) showed by combined scanning calorimetry and freeze-fracture electron microscopy that cytochrome oxidase can move laterally in the mitochondrial membrane. Determination of the mobility of the enzyme in the membrane may be of importance from different points of view. It may be asked whether mobility is essential for enzymic activity, whether mobility and transient contacts with other membrane proteins (ATP synthase?) might be essential in energy transduction. Determination of the enzyme's possible motion can also yield information about the size of the molecule in situ (monomer, dimer?).

Junge and his collaborators (Junge and De Vault, 1975; Kunze and Junge, 1977) determined the time-dependence of the linear dichroism following flash photolysis of the carbonmonoxy aa_3 complex in intact mitochondria. From these data they suggested that the rotational correlation time of cytochrome oxidase is at least 100 ms, implying almost full immobilization of the enzyme. However, more recent studies using the same technique, but with signal averaging and a better signal to noise resolution, have indicated a decay of anisotropy almost three orders of magnitude faster. This suggests fast rotation of cytochrome oxidase molecules, presumably around an axis perpendicular to the membrane, both in mitochondria and in reconstituted vesicles (Kawato *et al.*, 1980; Sigel, 1980). EPR studies using covalently attached spin-labels on subunits II and III (Swanson *et al.*, 1980), or on subunit III (Ariano and Azzi, 1980*a,b*), are consistent with rapid motion of a monomer or a dimer of the enzyme in reconstituted liposomes. The former workers also demonstrated that large aggregates of the enzyme may be formed in some purification procedures, exhibiting much slower motion, but with unaltered catalytic activity.

These studies show that cytochrome oxidase monomers or dimers (a distinction was not possible here) rotate with high velocity in the membrane. It seems clear that enzyme immobilization is not required for maximal activity.

VII. Conclusion

Mitochondrial cytochrome oxidase is probably composed of seven different subunits, of which the three largest (I, II and III) are coded by mitochondrial genes and translated inside the mitochondrion, while the rest are of cytoplasmic/nuclear origin.

The cytochrome aa_3 monomer (two haems plus two coppers) consists of one set of subunits in relative one-to-one molar stoicheiometry. However, the enzyme forms stable dimers, most notably in non-ionic detergents. $(aa_3)_2$ is also the unit present in lateral crystals derived by Triton X-100 extraction of mitochondria. There is no unequivocal evidence for fully active monomeric oxidase, and it is possible that the dimer is the form of the enzyme *in situ*.

The oxidase complex spans the membrane completely. Almost one-half of the enzyme mass protrudes into the aqueous space on the cytoplasmic surface, where cytochrome *c* binding occurs. The matrix side of the monomer is divided into two domains, giving it the shape of an asymmetric "Y", but little of the protein protrudes in the aqueous space on this side.

Labelling studies have identified three domains of the enzyme, and suggested location of subunits with respect to these domains. Thus some subunits, or portions thereof, are located in the aqueous C-domain (II, III, VII; possibly V, VI and I), some are in the aqueous M-domain (IV and possibly III and VII), while some are in contact with the phospholipid bilayer (I, II, III, VII and IV to some extent). Some subunits may themselves span the membrane (notably III and VII).

Subunit functions are more obscure, but there are indications for I and II being primarily involved in binding of the redox centres, and for III being involved in H^+ translocation (Chapter 7). Subunit II provides the binding site for cytochrome c.

The enzyme is highly mobile in the membrane, showing both lateral and rotational motion. No specific function of phospholipids is demonstrated other than for provision of a suitable two-dimensional "solvent". However, very tightly bound cardiolipin may be functionally important in providing the enzyme with a second binding site for cytochrome c.

4

Physical properties, configuration and topography of the redox centres

I. Introduction

Much work has been done since the remarkable studies by Keilin and Hartree with the microspectroscope. The haems and coppers of cytochrome oxidase have been studied in detail by a large variety of physical techniques. Even though this work has been most fruitful and has given us a comparatively detailed picture of the "active centres" of this enzyme, it has all along been hampered by problems of interpretation. This difficulty has mainly been caused by interactions between the redox centres (see Malmström, 1973), and dates back to the "classical" controversy between proponents of a "unitarian" and a "dualistic" view of the enzyme (see Lemberg, 1969). In fact, the problems were great enough to revive this controversy in a more modern form in the beginning of the 1970s (for reviews see Nicholls, 1974b; Erecińska and Wilson, 1978). Although there is general consensus today favouring the intrinsic difference between cytochromes a and a_3 in the enzyme, it should not be forgotten that the now well documented concept of haem/haem interactions stems from the "unitarian" hypothesis, in which spectroscopic and kinetic differences between haems a and a_3 were entirely ascribed to such interactions.

The "new wave" in the 1970s brought with it a new kind of controversy, namely with respect to the nature of the haem/haem interaction. One school considered it mainly to affect the spectroscopic properties, whereas the other school retained the original spectral definitions of cytochromes a and a_3 of Keilin and Hartree (1939), and ascribed the interactions to oxidoreduction effects (see Nicholls and Chance, 1974; Malmström, 1973; Wikström et al., 1976; Wilson and Erecińska, 1978). Since these different views include a very large difference in interpretation of the spectroscopic properties of cytochromes a and a_3, the controversy has had far-reaching consequences in the interpretation of nearly all functional data on the enzyme.

In this chapter we will discuss the properties of the individual redox centres, including their topography in the membranous enzyme. We will put considerable emphasis on the above-mentioned controversy due to its importance, and argue that it can, in fact, be solved in the light of present data. This conclusion is important also for the interpretation of kinetic data, as will be shown in Chapter 6. In Chapter 7 it will be suggested that the new information may yield a picture of cytochrome oxidase which borrows some essential features also from the now disproven "unitarian" hypothesis. It seems that cytochrome oxidase may provide us with yet another example of how Nature may be described better by a theory that is a compromise between rigid opposites.

II. Structure and physical properties of the redox centres

A. The $g = 3$ EPR signal and cytochrome a

EPR spectroscopy of the "resting" oxidized enzyme as isolated (Chapter 2), unperturbed by either extraneous ligands or by electrons, suggests *a priori* that the two haems must have very different molecular configurations. Thus one-half of total ferric haem is observed as a low spin signal with rhombic symmetry and typical g-values of 3, 2 and 1.5 (Van Gelder and Beinert, 1969; Hartzell and Beinert, 1974; Aasa *et al.*, 1976). This signal is reasonably homogeneous and also behaves as a single species in reductive kinetic studies (Chapter 6). It therefore seems safe to conclude that this set of signals (hereafter called the $g = 3$ signal) must stem from a single low spin ferric haem centre. Comparison of this signal with known haem compounds and haemoproteins indicates that it may be due to a *bis*-imidazole ferric haem (Babcock *et al.*, 1979; Peisach, 1978; Blumberg and Peisach, 1979). Theoretical simulations suggest that this haem centre is magnetically isolated, the distance from any other paramagnetic centre being at least 6–10 Å in the oxidized resting enzyme (Aasa *et al.*, 1976; Blumberg and Peisach, 1979).

Several lines of evidence point towards an assignment of the $g = 3$ signal to the haem of cytochrome a as originally proposed by Van Gelder and Beinert (1969). On reduction of the enzyme with dithionite or cytochrome c, cytochrome a is rapidly reduced (Chapter 6) and the $g = 3$ signal disappears (Hartzell *et al.*, 1973; Beinert *et al.*, 1976), while cytochrome a_3 reduction is very much slower. This evidence is not conclusive, however, since reduction of cytochrome a might induce a spin state transition in cytochrome a_3, which could be responsible for the loss of the $g = 3$ signal (Wilson and Leigh, 1972; Wilson *et al.*, 1976). On the other hand, the $g = 3$ signal persists unchanged in partially reduced oxidase complexes with either CO or NO bound to ferrous a_3 (Hartzell and Beinert, 1976; Wever *et al.*, 1974; Leigh *et al.*, 1974; Wilson *et al.*, 1976), strongly supporting the assignment of the $g = 3$ signal to ferric haem a.

Further support for this assignment comes from more recent magnetic circular dichroism and magnetic susceptibility studies (see below), which indicate that the oxidized "resting" enzyme contains one high spin and one low spin ferric haem, of which the latter is obviously responsible for the $g = 3$ signal. The assignment of the latter to cytochrome a is based on the definition that cytochrome a_3 is the haem centre that reacts with ligands such as O_2, CO, HCN, etc. We should stress, however, that there is an imminent danger in relying on ligand effects for spectroscopic assignments of the different redox centres. Ligand binding to one haem might cause a change in the physical properties of the other, confusing the assignment. It must therefore be considered fortunate that even though haem/haem interactions are indeed prominent in cytochrome oxidase, they are mainly restricted to redox interactions. Only small interactions are seen in the spectroscopic parameters. This simplifies interpretation of the spectroscopic data considerably. However, this is a thesis that due to its prime importance requires rigorous proof, and will therefore be discussed in a separate section (see Section V).

We conclude that the $g = 3$ signal in the fully oxidized "resting" enzyme can be assigned to ferricytochrome a, which is a low spin ($S = \frac{1}{2}$), magnetically isolated centre, possibly with imidazole ligands in both axial positions.

B. The haem a_3/Cu$_B$ centre

Van Gelder and Beinert (1969) originally suggested that the "quenching" of one of the two ferric haems (i.e. haem a_3, see above) from the EPR spectrum of the oxidized enzyme might be the result of antiferromagnetic coupling of this haem with cupric ion. Indeed, the $g = 2$ signal attributed to copper (Chapter 2 and Section II.C) accounts only for 40% of the functional copper present in the enzyme. This proposed spin coupling between haem a_3 and one of the coppers was substantiated theoretically by Griffith (1971) as a likely reason for the absence of EPR resonances from either centre.

Subsequent MCD experiments by Babcock et al. (1976) and Thomson et al. (1976) strongly supported this idea, indicating that the oxidized enzyme contains high spin ferric cytochrome a_3 and low spin ferric cytochrome a. More recent near-infrared CD and MCD studies have shown separate characteristic transitions for both ferric and ferrous a and a_3 (Eglinton et al., 1980). These may be very useful probes of the state of the individual haems.

More direct evidence for a coupled high spin ferric haem iron/cupric copper pair has come recently from studies of magnetic susceptibility at different temperatures (Falk et al., 1977; Tweedle et al., 1978; Moss et al.,

1978). These results indicate that the oxidized enzyme contains two magnetically isolated spin $S = \frac{1}{2}$ centres (ferric haem a and "visible" copper, or Cu_A^{II}, respectively, see Section II.C), and a spin-coupled $S = 2$ centre. The latter may best be interpreted as arising from high spin ferric cytochrome a_3 ($S = \frac{5}{2}$) antiferromagnetically coupled to Cu_B ($S = \frac{1}{2}$), yielding a binuclear centre of total spin $S = 2$. The temperature-dependence of the magnetic susceptibility indicates that the coupling between the metals is very strong (the absolute value of the exchange integral J is much larger than kT). A value of $-J \geq 200$ cm^{-1} has been implicated, which would restrict the number of possible chemical structures (see below).

Very strong magnetic coupling between haem iron and copper of a_3 and Cu_B in the oxidized enzyme would also suggest that the two metals must be very close to one another. A metal to metal distance of less than 7 Å is often quoted. This conclusion is substantiated by recent EPR studies using NO as ligand (Section III.B).

Seiter and Angelos (1980) recently suggested an alternative explanation to these data. In their view haem iron of a_3 may be in the quadrivalent ferryl (Fe^{IV}) state, and Cu_B in the cuprous state, in the fully oxidized "resting" enzyme. Cu_B was suggested not to undergo oxidoreduction during catalysis, but haem iron would instead effectively function as a two-electron carrier shuttling between the ferryl and ferrous states. However, several findings contradict this specific hypothesis. First, the optical spectrum of the oxidized enzyme shows no resemblance to haemoproteins with ferryl haem iron (see, for example, George, 1953; Yonetani, 1970). Second, recent Soret excitation Raman spectroscopy (Babcock et al., 1981) of the oxidized enzyme is indicative of a ferric state of haem a_3. Third, the long-sought-for EPR spectrum of Cu_B^{II} was recently detected (Section II.C) in a transient state of the enzyme, strongly corroborating the assumption that Cu_B undergoes oxidoreduction during catalysis. Finally, if haem a_3 were in the ferryl state in the oxidized enzyme, then potentiometric/spectroscopic titrations should reveal this, either as two spectroscopically typical one-electron transitions (ferryl $\rightarrow$ ferric $\rightarrow$ ferrous), or possibly as one two-electron event. None of this is observed, however (Chapter 5).

On the basis of this we favour the idea that haem a_3 is ferric and Cu_B cupric in the "resting" enzyme. However, haem a_3 might yet cycle through ferryl states in the aerobic catalytic cycle, but with simultaneous oxidoreduction of Cu_B (see Chapter 6).

Different models (see Fig. 4.1) for chemical structure of the haem a_3/Cu_B centre have recently been discussed (Reed and Landrum, 1979; Babcock et al., 1980). Palmer et al. (1976) suggested an imidazolate bridge between haem iron and copper on the "back" side of the haem (fifth co-ordination bond, the "front" side being that susceptible to attack to extraneous

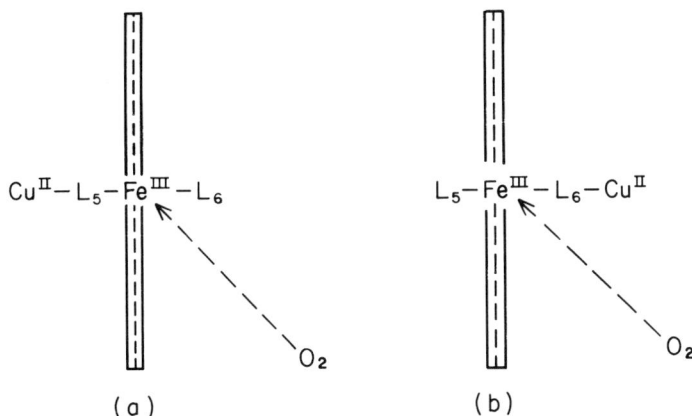

Fig. 4.1 Alternative structures of the haem a_3/Cu_B centre in the oxidized "resting" enzyme. Fifth ("back side") and sixth ("front side") axial ligands to ferric haem iron are denoted L_5 and L_6. The proposal by Palmer *et al.* (1976) corresponds to (a) with imidazolate and water as L_5 and L_6, respectively. In (b), L_6 may be oxygen (Blumberg and Peisach, 1979; Reed and Landrum, 1979) or carboxylate (Seiter *et al.*, 1978). L_5 may be either a histidine imidazole (Blumberg and Peisach, 1979), or may be vacant (Reed and Landrum, 1979). Arrow with O_2 shows the point of attack by extraneous ligands.

ligands). However, information from model compounds indicates that such a bridge may not mediate a strong enough magnetic coupling to fit the data on the a_3/Cu_B pair (Petty *et al.*, 1980; Reed and Landrum, 1979). In contrast, a μ-oxo bridge between iron and copper on the "front" side would fit this requirement, although other models (see, for example, Seiter *et al.*, 1978) should also be considered (Blumberg and Peisach, 1979; Reed and Landrum, 1979). The properties of model compounds suggest that in μ-oxo structures the haem iron is forced into a five-co-ordinate geometry (O'Keeffe *et al.*, 1975). This would contradict the μ-oxo structure if Babcock *et al.* (1981) are correct in concluding that the haem iron is six-coordinate in the oxidized enzyme, based on Raman spectroscopy. Shaw *et al.* (1981) recently measured the increase in the ^{18}O content of water after oxidizing the fully reduced enzyme with $^{18}O_2$ (i.e. one-half full turnover). All the reduced $^{18}O_2$ could be accounted for by $H_2^{18}O$ in the medium. These data contradict the idea that oxygen of O_2 is incorporated in a μ-oxo bridge of the oxidized enzyme, unless the bridge oxygen exchanges comparatively rapidly with solvent water.

It is clear that the final word has not been said on the structure of the a_3/Cu_B centre. The fifth ligand of haem a_3 is histidine (Section IV.C), as recently proved by Stevens and Chan (1981) by incorporation of ^{15}N-

labelled histidine into the yeast enzyme. It must be stressed, however, that this result was obtained with the reduced enzyme and need not necessarily apply on the fully oxidized enzyme.

We will in our discussion below (and see Chapter 6) use the μ-oxo model in the absence of a better proposal at the present time (but see also Note 5 on p. 191).

C. Spectroscopy and configuration of the coppers

Some of the essential properties of Cu_B have already been presented in the previous section. X-ray edge and EXAFS studies by Powers et al. (1979) have been interpreted to suggest that both coppers of the oxidized enzyme are cupric (contrast Seiter and Angelos, 1980; cf. above), and that Cu_B has a charge density similar to the Type I blue (Vänngård, 1972) copper protein stellacyanin, while the environment of Cu_A is more covalent.

Involvement of cupric Cu_B in the enzyme mechanism was recently strongly corroborated by the discovery of its long-sought-for EPR signal (Malmström et al., 1980; Reinhammar et al., 1980; Stevens et al., 1979a; Karlsson and Andréasson, 1981). This EPR signal was observed in a transient state of the enzyme following oxygenation (see Chapter 6).

The EPR spectrum of Cu_A (Chapter 2) is atypical of Cu^{II} in known copper proteins. Thus the g-value is very low, close to that of a free radical ($g = 2$), the signal usually appears to be devoid of hyperfine structure (but see below), and exhibits fast spin relaxation which prevents its detection at temperatures above about 150 K. For these reasons the assignment of the $g = 2$ signal to Cu^{II} has been questioned. Instead, it has been proposed that it arises from a free organic sulphur radical in close association with cuprous copper (Chan et al., 1978, 1979; Peisach, 1978; Blumberg and Peisach 1979; Hu et al., 1977).

More recently, Froncisz et al. (1979) detected hyperfine splitting of the EPR signal by low frequency EPR spectroscopy. Their results indicated hyperfine interactions with copper, plus another magnetic interaction, possibly with haem a. If the latter is correct, haem a may, after all, lie at a distance close to 10 Å from Cu_A and not be entirely isolated magnetically. Also, Greenaway et al. (1977) concluded from computer simulations of EPR spectra that Cu^{II} may interact with the low spin ferric haem iron (i.e. haem a in the oxidized enzyme), which may lie at a distance of about 7 Å from the copper. According to these workers the copper may indeed be cupric, but was suggested as having a peculiar electronic and stereochemical configuration that may explain the anomalous EPR spectrum.

The final word in this controversy was provided by Hoffman et al. (1980), who showed by ENDOR studies that Cu_A is in the Cu^{II} state in the oxidized enzyme, the magnetic properties being inconsistent with a thiyl radical co-ordinated to Cu^{I}.

The optical absorption of Cu_A has long been controversial and that of Cu_B most uncertain. It has now been established that the 820–840 nm band of the oxidized enzyme is due, at least to a proportion of 85%, to Cu_A^{II} (Tsudzuki and Wilson, 1971; Beinert et al., 1976, 1980; Boelens and Wever, 1980; Eglinton et al., 1980). With respect to interpretation of the kinetics of the enzyme, in particular, this conclusion is most important (Chapter 6).

A specific optical absorption band of Cu_B^{II} has not yet been detected with certainty. Chance and Leigh (1977) observed absorption changes near 740 nm that may be attributed to this species. If so, the absorption is smaller than the 830 nm band of Cu_A^{II} by a factor of three to four, and may thus have gone unnoticed in the work by Boelens and Wever (1980). However, the 740 nm band could also be ascribed to haem. Beinert et al. (1976) suggested that the 655 nm band of the fully oxidized enzyme may be due to ferric haem a_3 in its specific linkage to Cu_B^{II} (see Section II.B). Recently Karlsson and Andréasson (1981) showed that disappearance of the EPR signal due to Cu_B^{II} was accompanied by synchronous appearance of the 655 nm band. This rules out that the latter is due to Cu_B^{II}|alone, and supports the earlier conclusion by Beinert et al.

D. The reduced enzyme

The optical absorption characteristics of the reduced enzyme were summarized in Chapter 2. MCD and magnetic susceptibility studies show that one of the two haems is of high spin (Section II.B; see also Ehrenberg and Yonetani, 1961; Eglinton et al., 1980). This is almost certainly haem a_3, in agreement with earlier suggestions based on reactivity with ligands and the optical spectra (cf. Lemberg, 1969). The data show convincingly that haems a and a_3 are also in a very different environment in the reduced enzyme.

The high spin state of haem a_3 as opposed to low spin haem a would be expected to be reflected in the optical spectra. The classical interpretation of the spectrum of the reduced enzyme (Keilin and Hartree, 1939; Yonetani, 1960a; Vanneste, 1966; Lemberg, 1969) is indeed in excellent agreement with this picture (Section V). According to this view ferrous a_3 has a weak α-band and a high Soret/α-band ratio while haem a has a strong α-band and a much lower Soret/α-band ratio. This is indeed a typical difference between high and low spin ferrous haemoproteins (e.g. Williams, 1961).

A unique feature of ferrous cytochrome a is its split Soret band, best observed at low temperatures (Wilson, 1967; Gilmour et al., 1967).

Fully reduced cytochrome oxidase reveals no EPR resonances, which is expected for the diamagnetic cuprous copper and ferrous low spin haem a.

The paramagnetic high spin ferrous a_3 ($S = 2$) is invisible by EPR due to the integer S-value that usually prevents detection. Ferrous haem a_3 may be made detectable, however, in the presence of NO (Section IV.C).

Conformational differences between the oxidized and reduced enzyme are well documented (Urry et al., 1967; Cabral and Love, 1972).

E. Further information from Mössbauer and resonance Raman spectroscopy

Mössbauer spectroscopy of the oxidized enzyme also indicates the presence of high and low spin haem (Lang et al., 1974), in full agreement with the present views.

Extensive resonance Raman spectroscopy has recently been performed with different forms of cytochrome oxidase. These studies were first conducted with excitation at 441.6 nm (Salmeen et al., 1978a,b; Babcock and Salmeen, 1979; Babcock et al., 1979). More recently, excitation at lower wavelengths was found more suitable in studying high spin ferric haem a_3 (Babcock et al., 1981; Ondrias and Babcock, 1980).

The stretching vibration of the formyl group of haem A was observed for haem a_3 in both the high spin ferric and ferrous states, indicating conjugation of the free formyl with the π-electron system of the porphyrin. In these states the formyl residue is thus likely to lie in the plane of the porphyrin ring. In ferric haem a_3 this was independent of the spin state, but in ferrous a_3 the formyl stretch was lost on transition to the low spin state. Thus binding of O_2 to haem iron would lead to a change in formyl group geometry, which might be an important controlling factor of both ligand and redox affinity (Babcock and Salmeen, 1979; Babcock and Chang, 1979).

In model low spin ferric haem A compounds the carbonyl vibration was observed at 1670 cm^{-1} (Callahan and Babcock, 1981). However, in contrast to the property of haem a_3 (see above), the highest frequency band of ferric haem a was found at 1650 cm^{-1}. Also ferrous haem a showed no apparent formyl resonance. This suggests that the formyl of haem a may carry out a special function. A covalent interaction between the formyl and the protein, possibly a protonated Schiff's base, might account for these findings (Ondrias and Babcock, 1980; Babcock et al., 1981). Such a special function could be involvement in the proton translocating mechanism (Chapter 7).

The Raman spectra of cytochrome oxidase in different redox and ligand states were also compared with the model compounds haem A (Fe^{III}) chloride (high spin; five-co-ordinate), haem A (Fe^{III}) bis-dimethyl-sulphoxide (high spin; six-co-ordinate) and haem A (Fe^{III}) bis-(N-methyl-imidazole), which is low spin and six-co-ordinate (Babcock et al., 1981). It was concluded from this comparison that haem a is low spin and six-co-ordinate in both the ferrous and ferric state. Haem a_3 was suggested to be high spin ferric and six-co-ordinate in the fully oxidized "resting" state (cf. Section II.B), and five-co-ordinate high spin ferrous in the reduced enzyme.

With respect to the possible reactivity of the formyl group of haem a, it may be recalled that it was thought previously that the formyls of haem A were unreactive in the native enzyme, and that a Schiff's base is formed with a lysine residue only after denaturation at high pH (see Lemberg, 1969). However, Morrison and Horie (1964) reported that the formyl group of haem a_3 may become reducible by borohydride after ligand formation of iron with HCN. This interesting finding would deserve experimental pursuit.

III. Transient forms of the fully and partially oxidized enzyme

A. "Oxygenated" and "pulsed" oxidase

Okunuki and his collaborators (Sekuzu et al., 1959) found that aeration of reduced cytochrome oxidase produces a species (the "oxygenated" form) characterized by a red-shifted Soret band (to 428 nm) and a higher extinction of the band near 600 nm, relative to the "resting" enzyme (cf. Chapter 2). Later work, amply reviewed by Lemberg (1969), showed that this is not a truly oxygenated species (i.e. an enzyme–O_2 complex), but the name persists in the literature and will also be used here.

"Oxygenated" oxidase reverts to the "resting" state spontaneously in a reaction that is accelerated by cytochrome c, and is relatively stable only when the turnover of the enzyme is zero or very low.

There has been some confusion in the literature due to the observation of several "oxygenated" forms of the enzyme after pulsing the reduced oxidase with O_2 (Orii and King, 1972; Muijsers et al., 1971). Kinetically "early" and rather unstable species have thus gone under the same heading as the original "late" or fairly stable "oxygenated" species (see Lemberg, 1969; Nicholls and Chance, 1974). We would like to stress the danger of calling all these species "oxygenated", because they are presumably very different indeed (Chapter 6). We retain here the term "oxygenated" only for the species originally observed by the Japanese group, and for which there is some structural information (see below). This information should not be taken to apply on other so-called early oxygenated species (such as "Compounds" I and II of Orii and King, 1972, Chapter 6).

The "oxygenated" oxidase accepts four reducing equivalents just as does the "resting" enzyme (Williams et al., 1968; Tiesjema et al., 1972). Its

EPR spectrum is identical with that of the "resting" enzyme (Muijsers *et al.*, 1971), and so is its MCD spectrum (Babcock *et al.*, 1976). It has been suggested that the "oxygenated" form may be a conformational variant of the "resting" enzyme (Wharton and Gibson, 1968), a view that is substantiated by CD spectra, which are clearly different for the two forms (Myer, 1972; Muijsers *et al.*, 1971). Also the kinetic responsiveness is different, the "oxygenated" form being reduced more quickly by dithionite under anaerobic conditions, while the "resting" form exhibits a lag in the reduction of cytochrome a_3 (Lemberg and Gilmour, 1967). The "oxygenated" oxidase also reacts much faster with HCN than does the resting form (Brittain and Greenwood, 1976).

Recent Raman spectroscopy (Babcock *et al.*, 1981) has revealed a marked difference between these two forms. A 1572 cm^{-1} resonance characteristic of six-co-ordinate high spin ferric haem A (and of haem a_3 in the "resting" enzyme) was absent from the "oxygenated" form, which instead exhibited a band at 1590 cm^{-1}, considered to be a marker of low spin ferric haem A. Also the red-shifted Soret band of the "oxygenated" form has been taken as an indication of a low spin state of a_3. Yet the MCD data (Babcock *et al.*, 1976) suggested a high spin state. This discrepancy has not so far been fully explained (but see Babcock *et al.*, 1981, for one possibility). We look forward to model studies with ferryl haems, which could be a possible alternative configuration of the "oxygenated" species (Chapter 6).

The oxygen "pulsed" enzyme was originally defined only in kinetic terms (Antonini *et al.*, 1977; Wilson *et al.*, 1978). When the fully reduced enzyme is pulsed with O_2 in the presence of ferrocytochrome c, the lag in oxidation of the latter observed with the fully oxidized "resting" enzyme (Chapter 6) is partially or fully relieved. Thus some immediate product of the reaction with O_2 *in the presence of cytochrome c* is kinetically more potent in catalysing electron transfer between cytochromes a and a_3. The electron transfer between cytochromes c and a is hardly altered. Recently Brunori *et al.* (1979) obtained the "kinetic" spectrum of the "pulsed" enzyme, and showed that it has a higher extinction (by about 4 mM^{-1} cm^{-1} on an aa_3 basis) than the "resting" state in the 605 nm region (cf. Rosén *et al.*, 1977). The spectrum is thus rather similar to that of the "oxygenated" enzyme. The two species also show at least a gross kinetic resemblance to one another (see Lemberg and Gilmour, 1967, and above).

Brunori *et al.* (1979) demonstrated that "pulsed" oxidase forms by reaction of 1 mol of O_2 per aa_3 unit and, further, that it accepts four electrons in the conversion to the fully reduced form. As pointed out by the authors, this does not necessarily mean that the redox centres have the same valency as in the "resting" oxidase. Instead, it is likely that a partially reduced

O_2 molecule remains bound to the a_3/Cu_B centre. On the other hand, it is not certain that the "pulsed" state is homogeneous, i.e. with all aa_3 units in the same state. The kinetic heterogeneity in the oxidation of haem a (see Chapter 6) may well cause a difficulty in assigning a correct structure for the "pulsed" oxidase. In an analogous but probably not identical transient state described by Shaw et al. (1979; Section III.B), the $g = 3$ signal of ferric haem a had developed only to some 70% of its maximal intensity in the fully oxidized enzyme.

The kinetic status of the "pulsed" oxidase is dealt with further in Chapter 6.

B. Transient states exhibiting EPR signals from haem a_3

High spin ferric haem signals in the $g = 6$ region of the EPR spectrum were first recognized in partially reduced forms of the enzyme (see Van Gelder and Beinert, 1969, and Chapter 5). Their assignment to either haem a or haem a_3, or both, has all along been problematic, in part due to the existence of different signals in this region of both axial and rhombic symmetry, and in part due to slight but significant variations in the g-values under different conditions, but also because usually only small fractions of the total haem present could be ascribed to be responsible for these transitions.

More recently, high spin haem signals that account for nearly all of the haem a_3 present have been detected in transient states of the enzyme generated by anaerobic oxidation of the reduced enzyme with ferricyanide or porphyrexide (Shaw et al., 1978; Beinert and Shaw, 1977), and also on addition of NO to the "resting" enzyme as described further in Section IV.B.

Under all these conditions the $g = 3$ signal is fully developed so that the observed rhombic high spin signal can be attributed unambiguously to ferricytochrome a_3. This must mean that the antiferromagnetic coupling between ferric a_3 and Cu_B has been broken, as also shown by the absence of the 655 nm band (cf. Section II.B). The simplest explanation of this would be that Cu_B is reduced under these conditions, which is consistent with the fact that no EPR signal is observed from Cu_B^{II}. This view is strengthened by the recent finding (Section II.C) that Cu_B^{II} is indeed an EPR-detectable species, at least in some conditions.

High spin ferric haem a_3 is conspicuously reactive with ligands such as sulphide, cyanide and azide when generated by the oxidant-pulse technique of Shaw et al. (1978). This contrasts to the poor reactivity of the "resting" enzyme (Section IV.A). Ligand binding also results in the corresponding low spin ferric iron–ligand complexes, which are identifiable by

EPR spectroscopy (Shaw *et al.*, 1978). Furthermore, haem a_3 in this state is very sensitive to O_2 and CO, although these ligands are not expected to bind to ferric haem iron. Yet the high spin signal disappears rapidly on addition of O_2, and increases in rhombicity with CO (Shaw *et al.*, 1978). This may suggest that these gaseous ligands reach the haem "pocket" in close association with the iron's sixth ligand position (Section IV.C), and that occupancy of this "pocket" exerts an influence on the electronic state of the haem iron. In the light of the recent infrared studies by Alben *et al.* (1981), showing CO-binding to copper at low temperatures, the effects of O_2 and CO described above may result from binding to Cu_B^I. This would be consistent with the idea that Cu_B may be reduced in the transient state exhibiting the high spin ferric haem a_3 signal.

The conditions of the experiment in which the EPR-detectable high spin state of haem a_3 is formed are such that electrons are accepted by the added oxidant from the reduced enzyme, presumably via cytochrome a (see Shaw *et al.*, 1979). Although the used oxidants are very efficient thermodynamically (e.g. porphyrexide, $E_{m,7} = 725$ mV), the oxidation of Cu_B may be kinetically limited under such conditions. Notably, the life-time of the generated state is no longer than a few seconds.

Recently Shaw *et al.* (1979) identified another form of the enzyme which appeared within 2 ms at room temperature after reacting fully reduced cytochrome oxidase with O_2, and which decayed subsequently in a few seconds. This species is almost certainly identical with "early oxygenated" intermediates (see Section III.A and Orii and King, 1972) and its possible relevance in the catalytic mechanism is discussed in detail in Chapter 6. However, the spectroscopic properties of this compound are interesting, and may therefore deserve some comment here.

In the optical spectrum the Soret band is red-shifted with respect to the "resting" enzyme, while the α-band exhibits a broad maximum at 580 nm in the difference spectrum relative to the "resting" state. Furthermore, this compound has a fully developed 655 nm band (Chapter 2 and see above), a fully developed $g = 2$ resonance due to Cu_A^{II} and no rhombic $g = 6$ signals attributable to ferric high spin haem a_3. Most interestingly, this species exhibits a unique set of EPR signals, not hitherto detected in any state of the enzyme, centred at $g = 5$, 1.78 and 1.69. It may also prove significant (cf. Section III.A) that the $g = 3$ signal due to ferric haem a was developed only to some 70% of its intensity in the fully oxidized "resting" enzyme (Shaw *et al.*, 1979; Beinert *et al.*, 1979).

The unique EPR signals were tentatively interpreted to derive from a high spin ferric haem system ($S = \frac{5}{2}$), but with magnetic interaction with another paramagnetic centre leading to the split in the high-field resonances (Shaw *et al.*, 1979). These authors (and see Beinert *et al.*, 1979) listed several good arguments to support their contention that these signals arise from haem a_3. For instance, the EPR resonances were rapidly abolished by several ligands specific for ferric haem a_3.

Alternatively, the $g = 5$ EPR complex might arise from an $S = \frac{3}{2}$ system composed of ferryl (Fe^{IV}) haem iron and cupric copper, exhibiting anti-ferromagnetic coupling, as proposed in Chapter 6. However, more data are clearly required for an unambiguous interpretation. In particular, studies by MCD spectroscopy and determination of the magnetic susceptibility of this compound and its temperature dependence are needed.

We should stress that this species is not identical with the "oxygenated" form (Section III.A), but is a much more unstable compound. It is prob-ably closely related to "Compound II" described by Orii and King (1972; see Chapter 6).

IV. The binding of ligands to cytochrome oxidase

A. Ligand binding to partially and fully oxidized enzyme

It is remarkable that oxidized cytochrome oxidase in the "resting" state, as isolated, is conspicuously unreactive with most ligands such as cyanide or azide, which react readily with ferric cytochrome a_3 in turnover conditions or in the partially reduced enzyme (see Nicholls and Chance, 1974; Erecińska and Wilson, 1978).

Cyanide reacts with oxidized oxidase at a very slow rate (Antonini et al., 1971; Van Buuren et al., 1972) causing a spectral change which is consis-tent with a high to low spin transition in ferric haem a_3. Although genera-tion of a low spin a_3^{3+}–HCN complex is also substantiated by MCD spec-troscopy (Babcock et al., 1976), no low spin EPR signal corresponding to this complex is generated in the "resting" enzyme. However, in the pres-ence of reducing equivalents a $g = 3.58$ resonance is generated, typical of low spin cyanide complexes (DerVartanian et al., 1974; Shaw et al., 1978).

The magnetic susceptibility studies of Tweedle et al. (1978) provide an explanation for this dichotomy. Their data indicated that when cyanide is added to the oxidized enzyme a binuclear complex, viz. a_3^{3+}–HCN $(S = \frac{1}{2})$–Cu_B^{2+} $(S = \frac{1}{2})$, is formed (cf. Section II.B), with relatively weak antiferromagnetic coupling $(-J \simeq 40\ cm^{-1})$ between low spin iron and copper. This yields a diamagnetic ground state for the spin-coupled pair $(S = 0)$ below a temperature of about 40 K. Above this temperature the susceptibility was progressively enhanced consistent with a thermally ran-domized excited state $(S = 1)$ of the binuclear complex, with an exchange integral comparable to kT. Hence the cyanide complex would be undetect-able by EPR, but MCD spectroscopy, usually performed at high temperature where the complex is paramagnetic, would still reveal the cyanide compound (see also Babcock et al., 1976).

One plausible explanation for the poor reactivity of the oxidized "resting" enzyme with cyanide and other ligands would be a bridging μ-oxo ligand between haem iron and copper which may protect these metals from reacting with extraneous ligands. If so, it would be expected that reduction of Cu_B which would break the μ-oxo bridge should lead to higher reactivity of partially reduced oxidase with ligands. Higher reactivity indeed appears to follow partial reduction, and under such conditions the iron–ligand complex is no longer magnetically coupled to the copper, the former becoming detactable by EPR spectroscopy.

On the other hand, cyanide reacts much faster with cytochrome a_3 in intact mitochondria, the rate enhancement being of the order of 10^4 (Wilson et al., 1972a). Although one may argue that it is difficult to exclude the presence of reducing equivalents from aa_3 in intact mitochondria, the cyanide reactivity is unchanged even in the presence of a large excess of ferricyanide, which is known to effectively keep the aa_3 highly oxidized by establishing an electron sink at the level of cytochrome c. However, even under such conditions the turnover of cytochrome oxidase might be fast enough to prevent "relaxation" to the resting state.

The reaction of cyanide with the oxidized enzyme may occur in two steps (Antonini et al., 1971; Wilson et al., 1972a; Nicholls and Hildebrandt, 1978). If so, the first step involves binding of the ligand to a site near the haem (copper?, amino acid side chain?) because it is not associated with perturbation of the haem's d-electron system, being optically silent. The second rate-limiting step would then be that characterized by the high spin to low spin transition most easily followed in the Soret band. However, another alternative that also explains the cyanide binding kinetics is the existence of two different forms of the oxidized enzyme in thermal equilibrium ("resting" and "oxygenated"?), of which only the other ("oxygenated", cf. Section III.A) reacts with HCN. In this alternative, high cyanide reactivity need not be explained by partial reduction of the enzyme (contrast above), and would be consistent with the high reactivity of the "oxygenated" form (Brittain and Greenwood, 1976), which has been shown to contain four oxidizing equivalents—just as the "resting" state. On the other hand, partial reduction with a break of the putative μ-oxo structure could well constitute a second independent explanation for high cyanide reactivity. Thus the high reactivity in turnover conditions need not imply that the "oxygenated" form is necessarily of catalytic significance.

Wilson et al. (1972a) demonstrated that the cyanide reaction is strongly pH-dependent in intact mitochondria, with an apparent pK of 6.9 of a "regulating" acid/base group, the acidic form being reactive with the ligand. This might suggest that the "oxygenated"/"resting" equilibrium is a function of pH, but other explanations are, of course, also possible. We may speculate that isolation of the oxidase from its natural membranous environment may favour an "alkaline form" of the enzyme which is unreactive with cyanide.

Azide reacts comparatively quickly with the oxidized enzyme in contrast to cyanide. However, at low concentrations of the inhibitor (but substantial in relation to the concentration required for inhibition of turnover) only small spectral changes are observed, including a slight blue shift of the Soret band (Muijsers et al., 1968; Nicholls and Chance, 1974). At high concentrations the Soret band is red-shifted as with cyanide and sulphide (see below), but the change is of a smaller magnitude as compared to these latter inhibitors. However, in neither case is any EPR-dectable low spin or high spin haem species formed (Van Gelder and Beinert, 1969; Wilson and Leigh, 1972).

In partially reduced states of the enzyme the situation is dramatically different. Infrared spectroscopy now reveals iron-bound azide (Caughey et al., 1976) and EPR shows a characteristic species with $g = 2.9$ (Van Gelder and Beinert, 1969; Wilson and Leigh, 1972; Wever and Van Gelder, 1974), indicative of low spin ferric haem iron to which azide is bound. The redox potential behaviour of this species is interesting under equilibrium conditions, and is discussed in detail in Chapter 5. It may be noted here that both for azide and for cyanide it is the free acid (i.e. HN_3 and HCN) that is the inhibitory species, not the anion (Stannard and Horecker, 1948).

A special effect of azide, shared to some extent by sulphide and by alkyl sulphides (all inhibitory ligands of ferric cytochrome a_3; see Nicholls and Hildebrandt, 1978), is that the reduced minus oxidized difference spectrum of cytochrome a is shifted in the α- and Soret bands to the blue by a few nanometres (see also Wilson, 1967; Gilmour et al., 1967; Nicholls et al., 1976). This is an important indication of the presence of haem/haem interactions in the enzyme, a phenomenon that will be given much consideration below (Section V and Chapter 5).

Sulphide reacts readily with oxidized cytochrome oxidase yielding optical spectral changes similar to those induced by cyanide (Gilmour et al., 1967; Nicholls, 1975) indicative of a high spin to low spin transition in ferric a_3. This is confirmed by the generation of an EPR complex ($g = 2.54, 2.23, 1.87$; Wever et al., 1975; Wilson et al., 1976) typical of low spin ferric haem-sulphide compounds. The higher reactivity of sulphide as compared with cyanide may simply be due to the fact that the former also tends partially to reduce the enzyme.

Formate and fluoride (Nicholls and Chance, 1974; Nicholls, 1976) react with oxidized cytochrome oxidase to form a high spin a_3^{3+}–ligand complex (see also Babcock *et al.*, 1976), which is accompanied by a blue shift in the Soret region in contrast to the red shift that accompanies formation of (low spin) cyanide or sulphide complexes. In contrast to the case with azide, the difference spectrum of cytochrome *a* (reduced minus oxidized) is unaffected by formate (Nicholls, 1976). Formic acid is the inhibitory ligand.

B. The effect of NO on oxidized cytochrome oxidase

Stevens *et al.* (1979*a*) have recently presented most informative data on the reactivity of oxidized cytochrome oxidase with NO.

They found that upon incubation of the "resting" enzyme with NO a clear rhombic high spin signal (with g_x = 6.16 and g_y = 5.82, Fig 4.2(a)) appeared in the EPR spectrum, without any detectable change in the g = 3 and g = 2 signals of ferric cytochrome *a* and Cu_A^{2+}, respectively. This signal, which at a maximum accounted for 60% of one of the two haems, must consequently have arisen from high spin ferric cytochrome a_3. Moreover, this effect of NO is reversible and causes no detectable change in the optical spectrum of the enzyme, indicating that no change in redox state of the haems or Cu_A occurs.

This suggests that NO interacts with Cu_B^{2+}, breaking the antiferromagnetic coupling between the copper and the iron of a_3, so that the high spin ferric a_3 (Section II.B) now becomes detectable by EPR. Cu_B^{2+} remains undetectable due to spin coupling to the paramagnetic ($S = \frac{1}{2}$) NO ligand. A special feature of this effect is that it occurred only with certain oxidase preparations, whereas other preparations could be "activated" with fluoride to exhibit the NO reaction (Stevens *et al.*, 1979*a*). Thus fluoride may bind to ferric high spin a_3 and thereby induce a conformational change that enables the reaction between Cu_B^{2+} and NO. Apparently, these effects are very sensitive to the configuration of the a_3/Cu_B site. Thus the "oxygenated" form of the enzyme (Section III.A) failed to show the NO effect.

Addition of cyanide to the NO-treated enzyme abolished the NO-induced high spin signal. Only a small low spin g = 3.5 signal (10% ferric low spin cyanide complex) appeared, but no changes occurred in corresponding cytochrome a^{3+} or Cu_A signals (Stevens *et al.*, 1979*a*). Apparently cyanide causes formation of the spin-coupled a_3^{3+}—HCN—Cu_B^{2+} complex under these conditions (cf. above, Section IV.A), displacing the NO and exhibiting no EPR signals. The minor g = 3.5 resonance may be due to reduction of a small fraction of the Cu_B centres with a break in the paramagnetic interaction.

The addition of azide to the NO-treated enzyme caused a completely different set of effects (Stevens *et al.*, 1979*a*). Optical spectra clearly

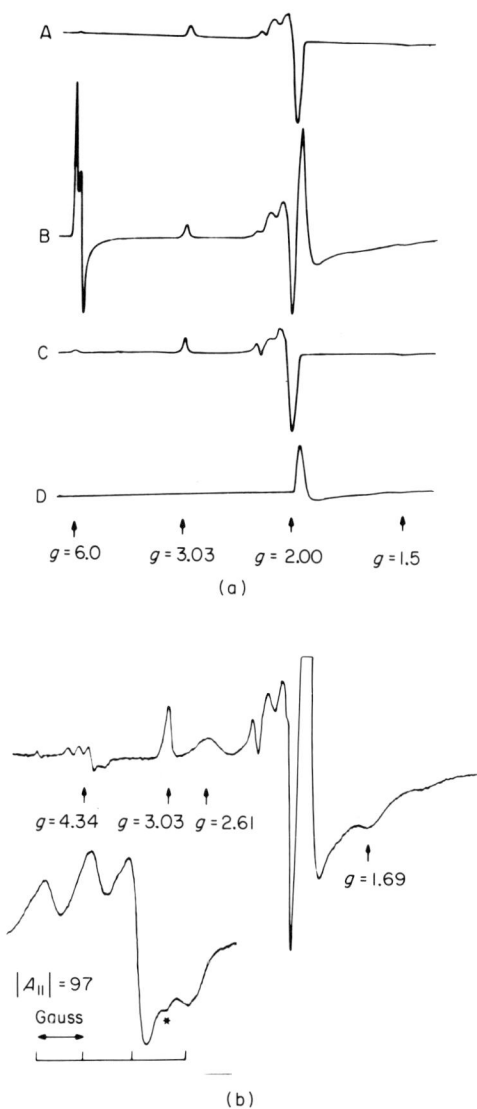

Fig. 4.2 The effect of NO (a) and N_3^- plus NO (b) on oxidized cytochrome oxidase. (a) Trace A, EPR spectrum of the "resting" enzyme; trace B, "resting" enzyme *plus* NO; trace C, NO removed from B; trace D, buffer *plus* NO. (b) EPR spectrum of "resting" oxidase treated with N_3^- *plus* NO. The inset in (b) shows the magnification of the half-field transition region. From Stevens *et al.* (1979a) with permission.

showed that under such conditions cytochrome a_3 is reduced and liganded to NO. No effects were observed in the $g = 3$ and $g = 2$ resonances, but the high spin $g = 6$ resonances disappeared. However, no EPR complex typical of a_3^{2+}–NO (Blokzijl-Homan and Van Gelder, 1971; Wilson et al., 1976; Barlow and Erecińska, 1979; Stevens et al., 1979b; and see Section IV.C) appeared. Instead, new EPR signals were observed at $g = 2$ and $g = 4.34$ consistent with a triplet state (Fig. 4.2(b)). The latter resonance exhibited a hyperfine structure, which was interpreted to be due to interaction with a copper ion of Type II (contrast Section II.C). These features were explained to be the result of formation of the complex Fe^{II}–NO–Cu^{II} with magnetic coupling of the unpaired electron on a_3^{2+}–NO with that of Cu_{BI}^{II} The system would thus be analogous to the µ-oxo structure (Section II.B), but with NO bridging between the two metals. The antiferromagnetically coupled spins of this $(S = 1)$ centre in the excited state exhibit a weak exchange interaction (about 5 cm^{-1}). This NO complex of cytochrome oxidase provides good support for the µ-oxo structure proposed earlier for the oxidized enzyme, and also supports ideas of "bridged" states in the catalytic cycle (Chapter 6).

The discriminative reduction of haem a_3 in the presence of NO and azide was explained as follows:

$$a_3^{3+} + NO + N_3^- \rightarrow a_3^{2+} + N_2O + N_2, \tag{4.1}$$

predicting the production of N_2O, which was also detected by mass spectroscopy (Stevens et al., 1979a; see also Brudwig et al., 1980).

C. Ligand binding to the reduced enzyme

The fully reduced enzyme readily binds CO, NO and HCN (for reviews see Lemberg, 1969; Nicholls and Chance, 1974). The binding of these ligands to ferrocytochrome a_3 results in a high spin to low spin transition which is revealed by MCD, magnetic susceptibility and optical measurements (Babcock et al., 1976; Tweedle et al., 1978; Ehrenberg and Yonetani, 1961; Thomson et al., 1976). In the optical spectra these ligands cause only a small decrease of the α-band at 605 nm which amounts to about 20–25% of the total reduced minus oxidized peak, and generate a new absorption peak between 585 and 600 nm (depending on the ligand) due to the a_3^{2+}–ligand complex (see, for example, Vanneste, 1966; Van Buuren et al., 1972). The effect of ligands on the optical spectra of cytochrome aa_3 is a very relevant problem with respect to optical assignment of the respective haems, and therefore for the interpretation of several different sets of data, notably potentiometric titrations. We therefore defer this particular subject to a separate section (Section V).

Treatment of reduced cytochrome oxidase with NO (or with hydroxyl-amine, which also produces the NO species) yields the species a_3^{2+}–NO, which has a characteristic EPR spectrum (Blokzijl-Homan and Van Gelder, 1971; Wilson et al., 1976; Barlow and Erecińska, 1979; Stevens et al., 1979b) with g-values of 2.09, 2.0 and 2.005. This interesting spectrum is due to the ligand's unpaired electron, which is distributed between iron and ligand. The latter resonance shows a strong hyperfine structure of nine lines, of which three lines have a large (21.1 gauss) and three lines a smaller (6.8 gauss) hyperfine splitting. The former can be ascribed to interaction with the nitrogen of NO, and the latter to interaction with the *trans* or fifth ligand of the haem iron (NO occupying the sixth position). Comparison of this compound with model haemoproteins of known structure liganded with NO indicates that the fifth ligand of haem a_3^{2+} is histidine. As ferrous a_3 is high spin in the reduced enzyme in the absence of extraneous ligands, it seems likely that the sixth position is unoccupied (cf. Babcock et al., 1980) or alternatively is occupied by a weak field ligand such as water.

Infrared spectroscopy of carbonmonoxy-cytochrome oxidase shows that CO is bound as a "terminal" ligand to haem iron, and that it is not bridged between Fe and Cu (Volpe et al., 1975; Caughey et al., 1976). The narrow band of the CO stretch shows that the ligand environment is ordered and isolated from the external medium.

Kinetic studies of photodissociation and reassociation of CO in reduced cytochrome oxidase at low temperatures (Sharrock and Yonetani, 1977) have revealed the likelihood of a "pocket" or binding site for CO (or O_2) near the haem iron with the capacity of only one ligand molecule and a large activation enthalpy for binding to the haem iron. The ligand appears to move into this "pocket" through a region mainly consisting of lipid. It therefore seems likely that dioxygen moves into the active site of cytochrome oxidase through the membrane and not directly from the aqueous phase. This is also in accordance with the fact that gaseous molecules such as CO and O_2 are much more soluble in a lipid milieu than in aqueous suspensions. The "pocket" near the sixth axial position of haem a_3 is hydrophobic, as indicated by infrared spectroscopy of the CO–enzyme complex (Yoshikawa et al., 1977), and also by the fact that it is the acid form of ligands such as azide and cyanide that inhibit the enzyme (Stannard and Horecker, 1948). Charged molecules do not seem to have easy access to this site. An essentially hydrophobic environment of haem a_3 is also supported by the Raman studies of Babcock et al. (1981), and by somewhat cruder "probe" studies (Section VI). The hydrophobicity of the haem a_3 vicinity may be of crucial importance for the mechanism of reduction of dioxygen, as pointed out by Yoshikawa et al. (1977).

Recent infrared data of Alben *et al.* (1981; see Chapter 6) suggest that the "pocket" or binding site of CO (and perhaps O_2) near haem a_3 may be Cu_B^I (cf. the findings regarding NO in Section IV.B).

V. Haem/haem interactions and interpretation of optical spectra

Studies of the function of cytochrome oxidase depend to a large extent on spectroscopic methods by which oxidoreduction, ligand binding, etc., are monitored for individual redox centres. No kinetic or thermodynamic description of the catalytic centres can be made until the spectroscopic signals can be unequivocally assigned.

Such assignments are, however, complicated by strong centre/centre interactions, of which many different kinds are possible in principle. Closely spaced centres such as haem a_3 and Cu_B are expected to exhibit short-range (direct) interactions. But various kinds of long-range interactions could easily be mediated by the protein structure (cf. haemoglobin). Furthermore, such interactions may involve any number of physical properties of the redox centres, e.g. optical spectra, magnetic properties, ligand binding, redox affinity, etc. In short, a very large number of possibilities exist and it may indeed be asked whether the present information is sufficient for a unique solution (Malmström, 1979; Lanne and Vänngård, 1978).

In this section we will argue that the large number of possible interactions may indeed be reduced to a minimum on the basis of the available information. We do not imply that all data can be immediately fitted to this solution (some data are, in fact, hampered by technical difficulties, making interpretation meaningless), but we wish to describe one principal solution which at least qualitatively—and in many instances quantitatively—fits the bulk of the experimental information.

We may here restrict ourselves to considering only haem/haem interactions. The haem/copper interaction between a_3 and Cu_B was discussed in detail above (Section II). Haem/copper interactions between haem a and Cu_A have not been described, with the possible exception of some magnetic effects (Section II.C). Haem/copper interactions between a_3 and Cu_A and haem a and Cu_B do not seem to occur, neither is there any clear evidence for copper/copper interactions of significance in the enzyme.

The haem/haem interaction in cytochrome oxidase is expressed in many interesting ways. One of the most fascinating modes of interaction is that of redox affinity. This is an anti-co-operative effect such that reduction of one haem dramatically reduces the affinity of the other haem for an electron. However, we will defer discussion of this effect to a separate chapter

(Chapter 5), because the only way to prove that such interactions indeed occur is to define first the spectroscopic properties of the redox centres.

A. Optical spectra of cytochromes a and a_3

The classical method of evaluating the individual spectra of the two haems is to make use of the fact that one (a_3) but not the other (a) binds ligands in both the ferrous and ferric forms. This procedure has been used extensively by several workers and the results are summarized in detail in the review by Lemberg (1969). Figure 4.3 shows the absolute and reduced minus oxidized difference spectra of cytochrome a and a_3 individually, along with their apparent molar absorptivities, as reported by Vanneste (1966) using this "ligand technique". As indicated in Section II, these results are at least qualitatively consistent with the known chemical difference between the two haems as revealed by magnetic susceptibility studies and by Mössbauer, MCD and Raman spectroscopy. However, and as also pointed out by Vanneste, these results are valid only under the assumption that the spectral properties of haems a and a_3 are independent. Thus, for example, the binding of cyanide to ferric cytochrome a_3 must not change the spectrum of cytochrome a and the redox state of cytochrome a must not affect the optical spectrum of the a_3^{3+}–HCN compound, etc. We should interject here that *small* interactions of this kind may be "allowed" for most purposes as they may be smoothed out or averaged by using several different ligands.

Another possibility for finding out the individual spectra would be to make use of an anticipated difference in redox affinity or midpoint redox potential between the two haems. Thus a potentiometric titration under equilibrium conditions combined with optical spectroscopy (see Wilson *et al.*, 1972b) may distinguish between high and low potential haems and reveal their respective spectroscopic properties. However, and just as with the "ligand method", this would lead to a simple determination of the spectral properties of a and a_3 only if no appreciable interactions occur, this time with respect to redox affinity (see Malmström, 1973).

As it turned out, the two techniques unfortunately yield very different results. In the redox titrations two species with half-reduction potentials differing by some 160 mV were indeed discerned, but with approximately equal spectral contributions to both the α- and the Soret bands (Wilson *et al.*, 1972b; Chapter 5 and contrast Fig. 4.3). Wilson *et al.* (1972b) assumed implicitly that no redox interactions occurred. The apparent discrepancy with the results of the "ligand technique" was instead explained as being due to a large increase in the molar absorptivity of reduced minus oxidized cytochrome a upon binding of ligands to either oxidized or reduced cyto-

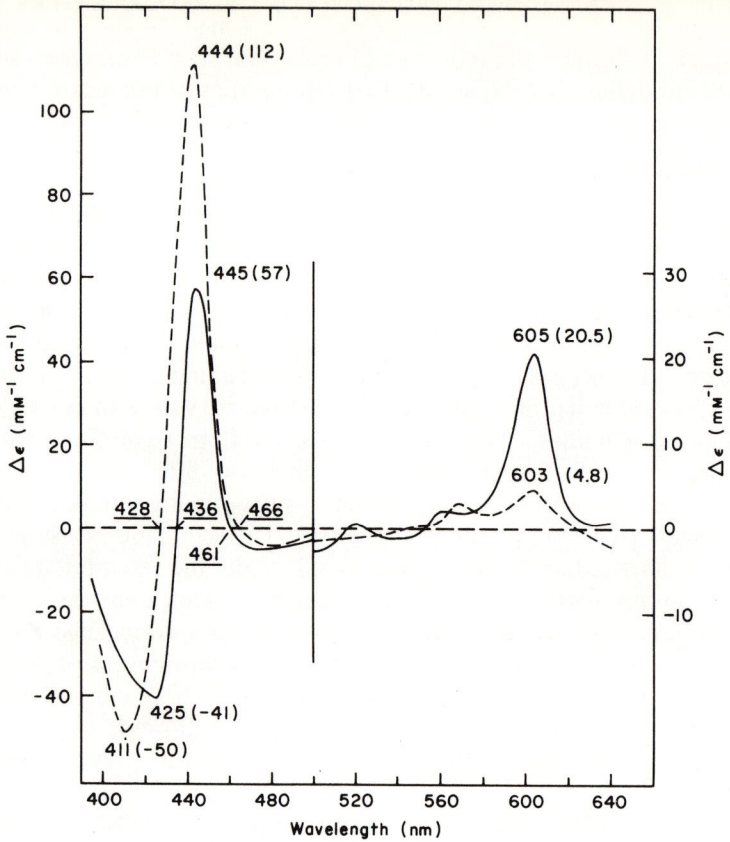

(a)

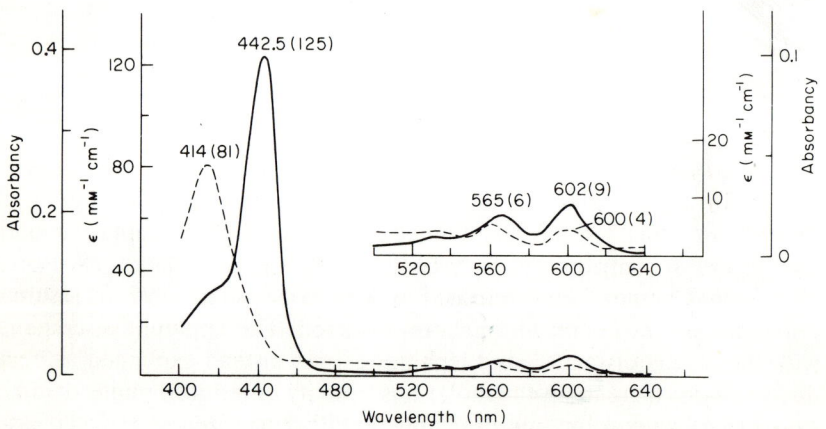

(b)

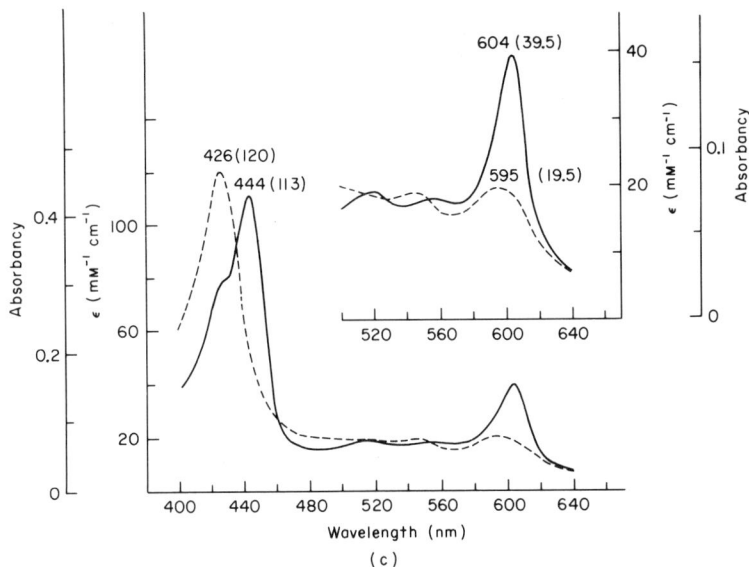

Fig. 4.3 Computed spectra of cytochromes a and a_3 by the "ligand method". (a) Difference spectra (reduced *minus* oxidized) of cytochromes a (——, i.e. $a^{2+} - a^{3+}$) and a_3 (– – – –, i.e. $a_3^{2+} - a_3^{3+}$). (b) Absolute spectra of reduced (——, i.e. a_3^{2+}) and oxidized (– – – –, i.e. a_3^{3+}) cytochrome a_3. (c) Absolute spectra of reduced (——, i.e. a^{2+}) and oxidized (– – – –, i.e. a^{3+}) cytochrome a. From Vanneste (1966) with permission.

chrome a_3. However, as pointed out by Malmström (1973) and by Nicholls (1974b), there was no experimental basis for this choice of interaction in optical but not in redox parameters. Wikström *et al.* (1976) showed subsequently that the optical/redox titration data could also be fitted quantitatively with or without ligands assuming no spectral interactions but with redox interactions instead (and see Chapter 5). A choice between these two primary models is of fundamental importance for the understanding of cytochrome oxidase function as discussed above. Of course, both kinds of interaction may occur, so that neither model may be strictly correct.

B. CO dissociation and the interpretation of optical spectra

The spectral haem/haem interaction model (Wilson *et al.*, 1972b and Section V.A) with non-interacting midpoint potentials of cytochrome a and a_3 can be tested experimentally for the postulated large increase in molar absorptivity of cytochrome a upon binding a ligand to cytochrome a_3. Wikström *et al.* (1976) proposed to carry out this test by comparing the difference spectra of CO photolysis for the fully reduced and half-reduced enzyme, respectively. In case there is a large change in molar absorptivity of cytochrome a on ligand binding or release, this should show up as a clear-cut difference between these two different spectra.

However, this test is complicated by two factors. First, there is uncertainty as to what extent the half-reduced photolysed species $a^{3+}a_3^{2+}$ may be converted to $a^{2+}a_3^{3+}$ due to intramolecular electron transfer. Second, at low temperatures photolysis of the "mixed" valence CO compound may easily be complicated by the generation of some "Compound C" (see Chapter 6) due to the presence of residual O_2 in the frozen sample. Thus it is not surprising that the photolysis difference spectrum of the half-reduced enzyme has been reported to be somewhat different in different research groups (Horie and Morrison, 1963; Greenwood et al., 1974; Wikström et al., 1976; Chance et al., 1979).

This important problem was therefore approached anew. After recording the familiar CO dissociation spectrum for the reduced enzyme (Fig. 4.4), the reduced CO–aa_3 preparation was treated with an excess of ferri-

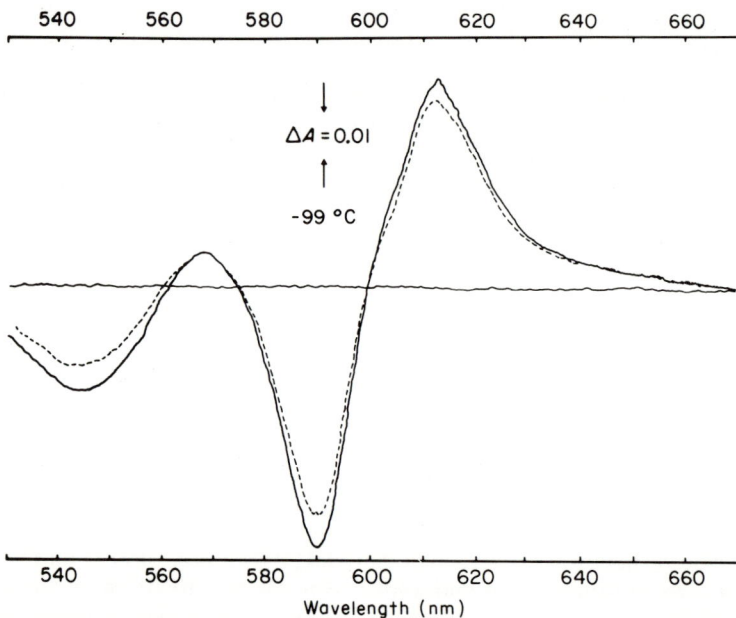

Fig. 4.4   CO photolysis difference spectra of fully reduced (———) and half-reduced (–––-) cytochrome oxidase. For details, see the text. From Wikström (1981b).

cyanide, under which conditions full oxidation of Cu_A and cytochrome a takes place (see, for example, Wever $et\ al.$, 1974). Only a minimum of adventitious O_2 was allowed to enter the preparation on ferricyanide addition. Photolysis of the preparation at $-100\ °C$ resulted in some formation of "Compound C", but this effect could be accounted for either by repetitive flashes and CO recombination, or by recording the difference spectrum of CO recombination rather than of CO photolysis.

With these procedures it was found that the CO dissociation spectra are identical within experimental error in the fully reduced and half-reduced states of cytochrome oxidase (Fig. 4.4, and see Horie and Morrison, 1963; Chance $et\ al.$, 1979).

The fact that the CO dissociation spectra are the same in the reduced and half-reduced enzyme shows that there is no significant change in the molar absorptivity of cytochrome a upon ligand binding/dissociation to cytochrome a_3. The spectral interaction model of Wilson $et\ al.$ (1972b) requires a 60–70% change in cytochrome a extinction, which would have been readily detected. Even the difference at room temperature reported by Greenwood $et\ al.$ (1974) is far from being large enough to satisfy this model, and is likely to be the result of some formation of "Compound C" in the half-reduced case, or to electron redistribution between haems a and a_3 following photolysis as described by Wever $et\ al.$ (1974). The spectra in Fig. 4.4 indicate that no such electron redistribution occurs at $-100\ °C$.

From these data we conclude that ligand binding to cytochrome a_3 does not cause any large changes in the optical spectrum of cytochrome a. Although this was proved only with CO, it is very likely to be true also for other ligands of reduced and oxidized cytochrome a_3 since all the known ligands have the same principal effect of abolishing only a small part (about 20–25%) of the enzyme's absorption peak at 605 nm in the reduced minus oxidized difference spectrum. However, this is not to say that there are no spectral interactions between the haems. In fact, some ligands of cytochrome a_3, notably azide, cause small spectral shifts in the absorption spectrum of ferrocytochrome a (Section IV.A, and see also Wikström $et\ al.$, 1976).

Several independent lines of evidence, besides the CO dissociation spectra, support the "classical" contention that the 605 nm band is mainly due to cytochrome a. This evidence is summarized in Table 4.1 (cf. Wikström, 1981b).

The importance of this conclusion cannot be overstated. It reduces considerably the number of possible interpretations of spectroscopic data both under equilibrium and under kinetic conditions (Chapters 5 and 6) and proves that strong negative redox interactions occur favouring the "neo-

Table 4.1   Evidence for the "classical" assignment of the main (about 80%) part of the 605 nm band of reduced cytochrome oxidase to cytochrome a.

1. CO dissociation/association spectra in reduced and half-reduced oxidase are identical
2. High extinction coefficient of 605 nm band on pulsing oxidized oxidase anaerobically with high concentrations of ferrocytochrome c (Wilson et al., 1975).
3. Identical Ca^{2+} shift of ferrocytochrome a spectrum at 605 nm whether cytochrome a_3 is oxidized or reduced, liganded or unliganded (Wikström et al., 1976).
4. No change in the α-band on Ca^{2+} shift in ferrocytochrome a_3 (Saari et al., 1980).
5. Correspondence of optical spectra with known high spin and low spin nature of haems a_3 and a, respectively.
6. Correlation of optical and MCD spectroscopy (Babcock et al., 1976, 1978).

classical" redox interaction model (see Nicholls, 1974b; Wikström et al., 1976). However, as pointed out by Malmström (1979), it is by no means obvious why such interactions should occur in the enzyme. In fact, it is not until very recently that these interactions may have acquired some functional significance, and then in a context not anticipated from the original neoclassical model (Chapter 7).

C. Other evidence for haem/haem interaction

The subject of haem/haem interaction in cytochrome oxidase has been much debated and studied, some aspects still being matters of controversy. The first school proposing such interactions was arguing for a "unitarian" model of cytochrome oxidase (for reviews see Lemberg, 1969; Nicholls and Chance, 1974; Erecińska and Wilson, 1978), according to which the two haems would be identical in the unperturbed enzyme, but would simulate the behaviour of two different haems due to such interactions, as a result of ligand binding or partial reduction. It is clear that this model can now be excluded (Malmström, 1973; Section II) in its original form. However, it is interesting from an historical perspective that these original ideas may still be valid in a modified form when applied to interactions between monomeric aa_3 units (Chapter 7).

Circular dichroism studies have been given contradictory interpretations. Myer (1971) concluded that no interactions were seen, but Tiesjema and Van Gelder (1974) found evidence for haem/haem interactions. As pointed out by Babcock et al. (1976), CD spectroscopy is, however, very sensitive to variations in the symmetry of the haem environment. Therefore, the rather small effects observed by Tiesjema and Van Gelder need not imply any substantial interaction with respect to the electronic structure of the haems themselves. In fact, Babcock *et al.* (1976, 1978) showed by careful MCD studies that the binding of ligands to cytochrome a_3, and/or redox changes in one haem, caused very small or no changes in the MCD parameters of the companion haem. However, while this is true with inhibitors such as cyanide and CO, it is a truth with modification for the case of sulphide and azide. In both cases the binding of the inhibitor to ferric a_3 causes a blue shift in the optical spectrum of cytochrome a (Section IV.A). However, these effects are small (though significant), and do not upset the more general conclusion of only small spectroscopic haem/haem interaction in the enzyme. In fact, the assignment of the main part of the α-band of cytochrome oxidase to cytochrome a (Section V.B) also implies that the reduced minus oxidized difference spectrum of this cytochrome may have a slightly different λ_{max} depending on whether cytochrome a_3 is reduced or oxidized (Wikström *et al.*, 1976), but again the difference is only 3–4 nm at the most.

Some aspects of proposed haem/haem interactions depend entirely on the adopted assignment of spectroscopic signals to specific redox centres. This is so, for example, for the effect of CO when added to half-reduced oxidase under anaerobic conditions, where the EPR signal at $g = 6$ (high spin ferric haem) is abolished and replaced by the low spin $g = 3$ signal. Leigh *et al.* (1974) interpreted this effect to be due to a high spin to low spin transition in ferric cytochrome a upon liganding ferrous a_3 with CO. However, this interpretation depends both on the assumption that cytochrome a_3 is of uniquely high potential and reduced before CO addition under such conditions (contrast Chapter 5) and on its corollary, viz. that the $g = 6$ signal stems from ferric cytochrome a. In the light of the data of Wever *et al.* (1974) and the necessity of adopting a scheme of strong redox interactions, these results may also be interpreted by the sequence

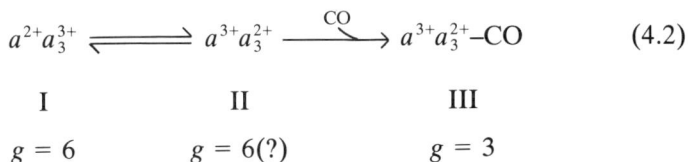

$$a^{2+}a_3^{3+} \rightleftharpoons a^{3+}a_3^{2+} \xrightarrow{\text{CO}} a^{3+}a_3^{2+}\text{–CO} \qquad (4.2)$$

I	II	III
$g = 6$	$g = 6(?)$	$g = 3$

Here the half-reduced enzyme contains the species I and II (Chapter 5), which give rise to the $g = 6$ signal(s). The main reason for the effect is therefore a shift of the redox equilibrium to the CO-bound state III in which ferric cytochrome a is low spin ferric. The phenomenon could also include a spin state transition as suggested by Wilson and Leigh (1972) provided that the half-reduced form of the enzyme in which haem a is oxidized and a_3 reduced (i.e. species II) indeed has haem a in the high spin state (Babcock *et al.*, 1978; Wikström *et al.*, 1976, and see Chapter 5).

It should be noted that most of the CO photodissociation experiments of Wever *et al.* (1974) were performed at room temperature, so that intramolecular electron transfer (from species II to species I in equation (4.2)) would appear entirely plausible. The fact that these workers observed no significant decrease in the $g = 3$ signal after photodissociation, but only production of the $g = 6$ signals could be attributed to either poor redox equilibration in their experimental system (only a stoicheiometric amount of NADH was added, see Chapter 5), or to appreciable further oxidation of the species I (see equation (4.2)) to $a^{3+}a_3^{3+}$ in experiments containing an excess of ferricyanide.

By contrast, the experiments performed by Leigh *et al.* (1974) were done under carefully redox-poised conditions. These workers observed, besides the generation of the $g = 6$ signal(s), an appreciable decrease in the $g = 3$ signal on CO photolysis of the half-reduced enzyme. A most interesting finding was that these effects persisted, though to a smaller extent, when the sample was illuminated at very low temperatures, down to 5–10 K. Since intramolecular electron transfer between haems a and a_3 is unlikely to occur at such low temperatures (see Fig. 4.4, Section V.B, showing that it does not occur at $-100\ ^\circ$C), it may be necessary to interpret a portion of the EPR changes on CO photolysis of the half-reduced enzyme as being due to true haem/haem interaction in which the binding of CO to $a^{3+}a_3^{2+}$ causes a transition of the spin state of cytochrome a^{3+} from high to low spin. It is most remarkable that this effect occurs within 5 ms at 5 K (Leigh *et al.*, 1974), indicating that the haems may be close to one another in the half-reduced form of the enzyme. The only weakness of these interesting results is that the observed changes in the EPR spectrum only accounted for a small fraction of the oxidase molecules present. However, in some experiments with the isolated enzyme the reactive fraction was as high as 30%.

Another interesting expression of the haem/haem interaction comes from the work of Wilson *et al.* (1976) on the EPR properties of the a_3^{2+}–NO species. This species is uniquely detectable by EPR (see Section IV.C), but the EPR signals are shifted slightly but significantly upon oxidoreduction of cytochrome a (and Cu_A). A careful study of the $g = 3$ signal in the species $a^{3+}a_3^{2+}$–NO would be of interest to find out whether the paramagnetic a_3^{2+}–NO centre might affect the haem a magnetically, which, if so, could provide an estimate of the distance between the two haems in the half-reduced enzyme.

The influence of the redox state of cytochrome a on haem a_3 is also revealed in the different affinity of $a^{2+}a_3^{2+}$ and $a^{3+}a_3^{2+}$ for CO (Greenwood *et al.*, 1974), which is also reflected in the midpoint potentials (Chapter 5). However, the redox state of haem a apparently has no influence on the geometry of the formyl group on haem a_3 (Babcock *et al.*, 1981).

There seems to be little doubt that haem/haem interactions exist in the enzyme. These interactions have only small but significant effects on the spectroscopic parameters. The main interaction appears to be one that affects the redox parameters (Chapter 5).

VI. Topography of the redox centres

Until very recently it has not been possible to locate the redox centres to any particular subunit. Previous proposals in this regard have been based on comparatively weak premises, and even today the evidence is not conclusive (see Chapter 3). The best current model seems to be one where subunits I and II are primarily involved in the "anchoring" of the haems and the coppers. It is even possible that subunit I *is* cytochrome a_3 and subunit II *is* cytochrome a, but it must be remembred that the redox centres might be "sandwiched" between two or more polypeptide chains, each of which might contribute with hydrogen-bonding residues, hydrophobic groups and ligand moieties (see Chapter 3).

As discussed in Chapter 3, a large proportion of the enzyme's mass protrudes into the aqueous C-phase, which also includes the cytochrome c-binding domains. Fluorescence resonance transfer studies (Vanderkooi *et al.*, 1977; Dockter *et al.*, 1978) indicate a distance of c. 25 Å between the haem of bound cytochrome c and the nearest haem A. This may be compared with the more recent studies of Ohnishi *et al.* (1979), in which magnetic interactions were measured between the water-soluble paramagnetic dysprosium–EDTA complex and the haems of the oxidase. These results suggest that haem a is situated fairly close to the C side of the membrane, which also appeared to be the locus of Cu_A. In contrast, haem a_3 interacted with the probe only in the isolated enzyme. In the membranous state no effect was found whether the probe approached the enzyme from the C or the M side of the membrane. Specific interaction of Ca^{2+}, probably with the propionate carboxyls of haem a, supports the C-side positioning of this haem (Saari *et al.*, 1980; Wikström and Saari, 1975). Ca^{2+} was also found to affect haem a_3 in an analogous fashion, but again only in the isolated state of the enzyme (Saari *et al.*, 1980).

The picture obtained from these results seems consistent, and agrees with the proposal by Winter *et al.* (1980) that subunit II (which binds cytochrome c and is exposed on the C side; Chapter 3) may bind haem a. It also agrees with the finding (see Chapter 6) that haem a is the first electron acceptor of the enzyme. In contrast, haem a_3 is apparently more buried, and might in part be shielded by the phospholipid bilayer. This is consistent with the proposal by Winter *et al.* (1980) that subunit I (which appears to be buried, cf. Chapter 3) might be specifically associated with haem a_3.

So far there are no reliable distance measurements between haems a and a_3. The absence of magnetic interaction puts the minimum distance at about 10 Å (Malmström, 1979). On the other hand, the haem/haem interaction that apparently takes place even at very low temperatures (Leigh et al., 1974), together with the fast electron transfer between the haems that seems to occur in some configurations even at low temperatures (Chapter 6), suggest that this distance may be no longer than about 25 Å (see Chance et al., 1977).

Finally, it should be remembered that the available distance measurements have been made only in specific states of the enzyme, for which they apply. A change in redox state, for instance, could cause changes in position of the redox centres with respect to one another. Information on such changes is lacking, however.

Studies with orientated membrane multilayers containing cytochrome oxidase have yielded important information on the orientation of the haems with respect to the membrane (Erecińska et al., 1977, 1978a,b; Blasie et al., 1978; Blum et al., 1978). These experiments have been performed with cytochrome oxidase vesicles as well as with mitochondrial membranes, essentially with the same results. Optical and EPR measurements have shown that both haems are orientated with their planes almost exactly perpendicular to the plane of the membrane (see Fig. 4.5). This

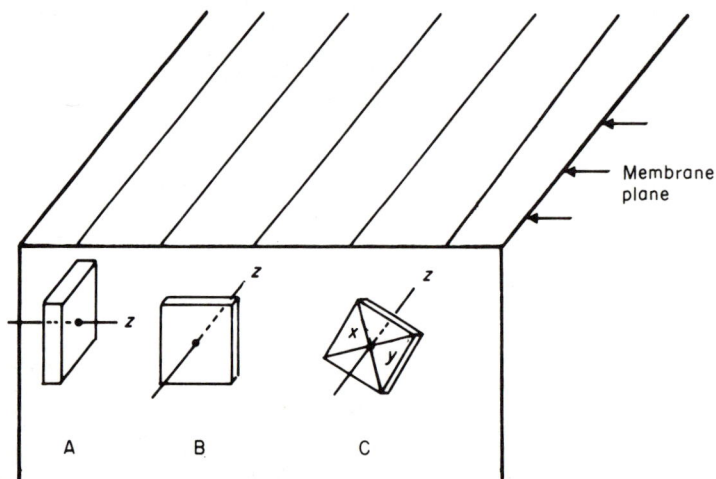

Fig. 4.5   Possible orientations of the haem disk with respect to the membrane. The z-axis perpendicular to the haem plane is shown in all three cases. Case C also shows the x- and y-axes within the haem plane. In all three cases the haem plane is perpendicular to the plane of the membrane (i.e. the haem normal, z, lies within the membrane plane).

orientation is also observed for the cytochrome oxidase haems in the organisms *Paracoccus denitrificans* and *Tetrahymena pyriformis* (Kilpatrick and Erecińska, 1977), as well as for the *b* cytochrome in the respiratory chain. In contrast, the haem of cytochrome *c* is orientated such that the normal of the haem plane is tilted about 70° from the normal of the membrane plane (Vanderkooi *et al.*, 1977; Erecińska *et al.* 1978*b*).

These data still do not define the orientation of the haems with respect to (i) rotation of the haem normal in the membrane plane, and (ii) rotation of the haem disk around its normal (i.e. the *z*-axis; see Fig. 4.5). Lack of optical dichroism of the haem spectra when the incident light is directed perpendicularly to the membrane plane indicates, in fact, that the haem planes are randomly orientated about an axis normal to the membrane (Erecińska *et al.*, 1978*a*). However, recent experiments by Blum *et al.* (1978) and by Erecińska *et al.* (1979) provide indications on the orientation of the haems around their normal (case (ii) above; cf. Fig. 4.5), although a definite assignment of the g_x- and g_y-axes with respect to the haem's nitrogen atoms and porphyrin ligands is somewhat ambiguous. But since it is reasonable to assign the g_y-axis to the axis through pyrrol rings with the highest electron-withdrawing power (Taylor, 1977), this might be identified with the N–Fe–N axis through vinyl- and formyl-bearing rings (see Fig. 2.1) in haem A. On this assumption the orientation of the g_x- and g_y-axes with respect to the membrane plane, as determined by EPR, can be interpreted in terms of the corresponding orientation of the *x*- and *y*-axes of the haem (see Fig. 4.5, case C).

In the oxidized enzyme the *g*-tensors of the $g = 3$ signal (i.e. haem a^{3+}; Section II.A) are orientated such that the *x*-axis is 30° from the membrane plane (the *y*-axis is correspondingly 60° from this plane; Blum *et al.*, 1978). This was confirmed by Erecińska *et al.* (1979) for the mammalian enzyme, but was found to be different in *Para. denitrificans* (*x*-axis 40°, *y*-axis 50° from membrane plane) and in *T. pyriformis* (0° and 90°, respectively).

In contrast, the EPR signals from low spin ferric haem a_3 in all three species indicated that it may be orientated similarly. The g_x-axis was found to be either 0° or 90° from the membrane plane with azide and sulphide as ligand, respectively. However, this difference probably reflects a rotation of the *g*-tensor axes by 90° in the haem plane rather than rotation of the haem in its own plane, as a function of the axial ligand.

Since the low spin ferric haem signals from *a* and a_3 are well defined in the sulphide- or azide-liganded enzyme (see previous sections), the above orientation data also adds to the now impressive amount of information on a different milieu around the two haems in cytochrome oxidase.

The relative orientations of redox centres as well as the distances between them are most relevant for the elucidation of mechanisms of electron transfer. So far the data is not sufficient to permit any profound conclusions and more work is required in this important area. At present two main findings might be emphasized, viz. (i) that the distance between the haems of cytochromes c and a, and possibly between haems a and a_3 as well, is quite long (25 Å and ≥ 10 Å, respectively), and (ii) that the two haems A are orientated perpendicular to the plane of the membrane. A distance of 25 Å is well within the distances expected for quantum-mechanical electron tunnelling as treated by De Vault and Chance (1966), but somewhat longer than expected from the treatment by Hopfield (1974; see also Chance *et al.*, 1977). The orientation of the haems might also be of significance with respect to the mechanism of proton translocation. It would easily allow a direct participation of the axial ligands in this process.

VII. Conclusion

Figure 4.6 may be more schematic than realistic, but the gross shape of the aa_3 monomer (Chapter 3) and its positioning in the membrane are approximately correct.

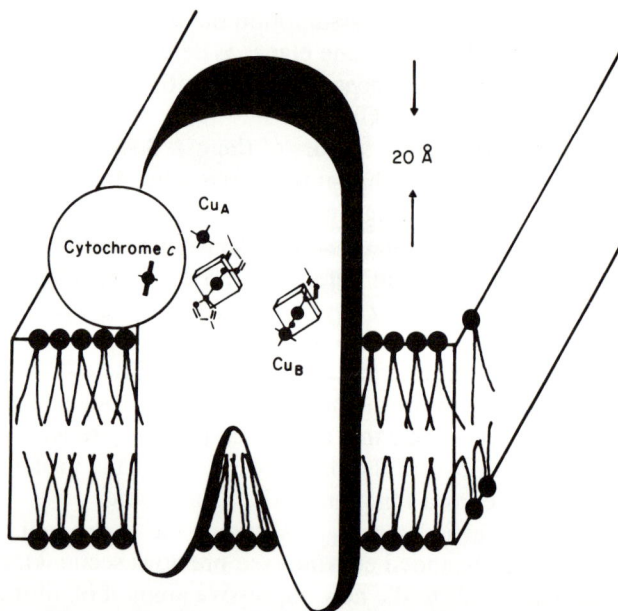

Fig. 4.6 Schematic representation of the gross structure and topography of the redox centres in a membranous monomer of cytochrome oxidase. For details, see the text and Chapter 3.

The four redox centres of the monomer are all different from one another. Haem a is (at least in most states) a six-coordinate low spin haem A (possibly with imidazole ligands in the fifth and sixth axial positions). Ferrous haem a_3 is high spin when unliganded, possibly five-co-ordinate, and with a histidine imidazole in the fifth axial position (at least in the low spin state with an extraneous sixth ligand). In the oxidized enzyme ferric haem a_3 is of high spin and possibly six-co-ordinate.

At least in the oxidized enzyme haem a_3 is located in very close proximity to Cu_B, the two metals being almost certainly coupled by a bridging ligand. The preferable structure might be a μ-oxo bridge in the sixth axial position of the haem, but other possibilities have not been excluded. The a_3/Cu_B centre lies in a hydrophobic region of the protein. In the membranous enzyme it is not near either aqueous phase on the two sides of the membrane. Haem a and Cu_A are probably located nearer to the C side of the membrane. Cu_A is bound to the protein in a geometry that is uncommon for copper proteins. The haem planes of both a and a_3 are oriented perpendicular to the plane of the membrane.

There are interactions between the haems in the enzyme, but these are only weakly reflected in spectroscopic parameters. The natures and mechanisms of these interactions remain obscure.

5
Oxidoreduction properties of the redox centres

I. Introduction

The thermodynamic properties of the enzyme's redox centres are of great importance for the full understanding both of catalysis and of energy conservation (which, indeed, is part of catalysis). Elucidation of these properties, like so much of the other information on this enzyme, has been impeded by the problems of identifying spectroscopic signals with specific centres (for useful discussions on this problem, see Malmström, 1973; Nicholls, 1974b; Nicholls and Chance, 1974; Babcock $et\ al.$, 1978; Lanne and Vänngård, 1978). A final resolution of the spectral contributions of haems a and a_3 to the 605 nm band of the fully reduced enzyme has now been obtained (Wikström $et\ al.$, 1976; Wikström, 1981b; Chapter 4), which provides a basis for the deconvolution of the identity of other spectroscopic signals. In addition, it strongly supports the contention that the haem/haem interactions in the enzyme are mainly of the type affecting the oxidoreduction parameter E_m.

In the following sections we discuss redox potential titrations of the enzyme under equilibrium conditions using both EPR and optical spectroscopy, and analyse the effects of ligands, including the proton, on the equilibrium redox parameters. In the main, this analysis will be restricted to the classical contention of the functional unit of cytochrome oxidase, viz. the monomeric cytochrome aa_3 molecule, containing two different haems and two different coppers. However, as already mentioned in passing (Chapter 3), we may have to consider a model in which the catalytic unit is the $(aa_3)_2$ dimer. The implications of this with respect to haem/haem interaction will not be considered in this chapter. To what extent conclusions in this chapter may have to be revised for a dimeric system will be discussed further in Chapter 7.

II. Redox titrations and their interpretation

A. The requirement of redox equilibrium

Cytochrome oxidase has been studied by a variety of redox titration techniques. All have in common the implicit assumption of redox equilibrium in the system. This is a very stringent requirement, meaning that all four redox centres experience the same redox potential (E_h) at all times, so that perturbations, were they imposed by a change in E_h or by addition of a ligand, do not change this situation in the time domain in which the measurement is made. Note that the redox equilibrium must be achieved both among the centres in individual molecules, as well as between different cytochrome oxidase molecules.

Unfortunately, it turns out that redox equilibrium is quite difficult to achieve under many experimental conditions with this enzyme. One reason for this may be the difficulties involved in lowering the O_2 concentration sufficiently to true "anaerobic" levels. A second reason may be the difficulty of achieving redox equilibrium among individual aa_3 units, for which the presence of cytochrome c seems to have a beneficial effect (see below). From a careful inspection of the literature it seems to us that the most reliable technique is the anaerobic potentiometry introduced for the respiratory and photosynthetic electron transport systems by Dutton and his collaborators (Dutton, 1971; Wilson et $al.$, 1972b; Dutton and Wilson, 1974).

Titrations with stoicheiometric amounts of NADH in the presence or absence of phenazine methosulphate (Van Gelder and Slater, 1963) were performed in pioneer work, in which the number of electron acceptors, their individual stoicheiometry and their extinction coefficients were determined for the enzyme. However, later work has revealed that this method often does not lead to redox equilibrium between the redox centres. In some cases the deviation from equilibrium may be slight (see, for example, Babcock et $al.$, 1978), but in others it may be considerable. From the by now generally accepted E_m values of the haem and copper redox transitions (see Tsudzuki and Wilson, 1971; Wilson et $al.$, 1972b; Tiesjema et $al.$, 1973; Lindsay et $al.$, 1975; Andersson et $al.$, 1976), it can be calculated that Cu_A is, for instance, much too extensively reduced in some experiments using this technique (see, for example Wever et $al.$, 1977).

Muijsers et $al.$ (1972) found only a single haem redox transition with E_m at 280 mV in the isolated enzyme titrated in the absence of cytochrome c. This is another example of the lack of redox equilibrium, in this case apparently due to the lack of cytochrome c (see Leigh et $al.$, 1974; Hartzell and Beinert, 1976; Schroedl and Hartzell, 1977a,b,c). Cytochrome c may catalyse electron equilibration among individual aa_3 units and might have additional effects on electron equilibration within the oxidase molecule (Krab and Slater, 1979; Chapter 7).

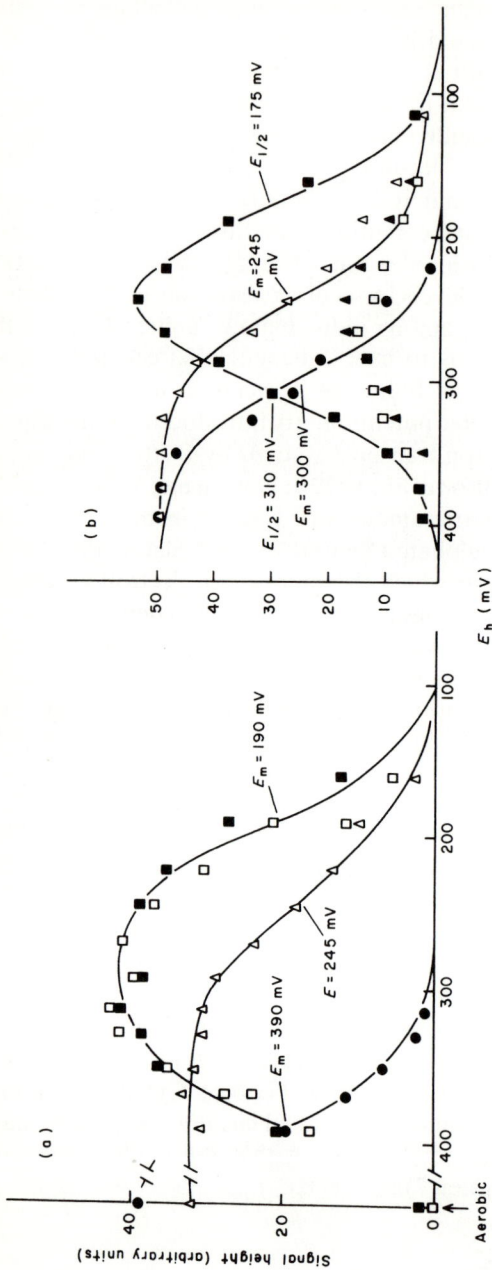

Fig. 5.1 The oxidoreduction potential dependence of the EPR signals of cytochrome oxidase in submitochondrial particles. Signal amplitudes are plotted as a function of the E_h value of the sample before freezing. (a) pH 7.0: □—□, $g = 6.4$ maximum; ■—■, $g = 6.03$ maximum, ●—●, $g = 3$ maximum; △—△, the copper (CuA) signal. (b) pH 8.5: ■—■, $g = 6$ maximum; ●—●, $g = 3$ maximum; △—△, the copper (CuA) signal; □—□, low spin haem signal at $g = 2.63$; ▲—▲, low spin haem signal at $g = 2.57$. $E_{1/2}$ values indicated are the potentials at which the $g = 6$ signal peak height has dropped to half its maximum observed value. This value differs slightly from the true midpoint potential. From Wilson et al. (1976) with permission.

B. Redox potential dependence of EPR resonances at equilibrium

Figure 5.1 shows potentiometric titrations of the cytochrome aa_3 system by Wilson et al. (1976) in submitochondrial particles, monitoring the EPR resonances. Similar results have been obtained subsequently with the isolated enzyme provided that extreme care has been taken to accomplish redox equilibrium. In general, we have the impression that equilibrium may be easier to accomplish in the intact membranes.

Three kinds of EPR signal are observed, viz. low spin and high spin ferric haem signals, and the $g = 2$ signal of Cu_A^{II} (Chapter 4). The last-mentioned behaves as a simple one-electron acceptor with $E_{m,7}$ of about 245 mV (Fig. 5.1(a)), independent of pH (Fig. 5.1(b)). Very similar data have been reported from titrations of the 830 nm band (Erecińska et al., 1971). No signal attributable to Cu_B^{II} has been observed in equilibrium conditions (cf. Chapter 4).

At pH 7 the $g = 3$ signal (with its associated high-field resonances; see Chapter 4) is the only low spin haem signal of the aa_3 system observed in mitochondrial membranes. This signal, which is due to ferric haem a, disappears monotonically on decreasing E_h as if it belonged to a haem centre with E_m between 300 and 400 mV depending on pH (Fig. 5.1).

During partial reduction of the enzyme, at E_h values between 300 and 400 mV depending on pH, new EPR signals appear that are absent both from the fully oxidized and the fully reduced enzyme. At pH 7 these are high spin ferric haem signals ($g = 6$), composed of at least two components with axial and rhombic symmetry, respectively. At high pH, however, new low spin haem signals ($g = 2.6$) partially replace the $g = 6$ signals, exhibiting a similar dependence on E_h (Fig. 5.1(b)). In the isolated enzyme, the latter signals are also observed at neutral ambient pH (see also Van Gelder and Beinert, 1969; Lanne et al., 1979). Taken together, the $g = 6$ and $g = 2.6$ signals account for between 25% (Hartzell and Beinert, 1976) and 50% (Wilson et al., 1976) of total haem at their maximum intensity (cf. also Lanne et al., 1979).

C. Redox potential dependence and interpretation of optical spectra

Figure 5.2 shows a redox potential titration of the optical absorption change at 605–630 nm (α-band) of cytochrome oxidase in mitochondria. Again, very similar data are obtained with the isolated enzyme under controlled equilibrium conditions (Tiesjema et al., 1973).

Figure 5.2(a) illustrates how the 605 nm band titrates as if two redox components were contributing about equally to the absorption band, but with widely different E_m values. Mathematical resolution of the two transitions into two one-electron couples is shown in Fig. 5.2(b). Very similar

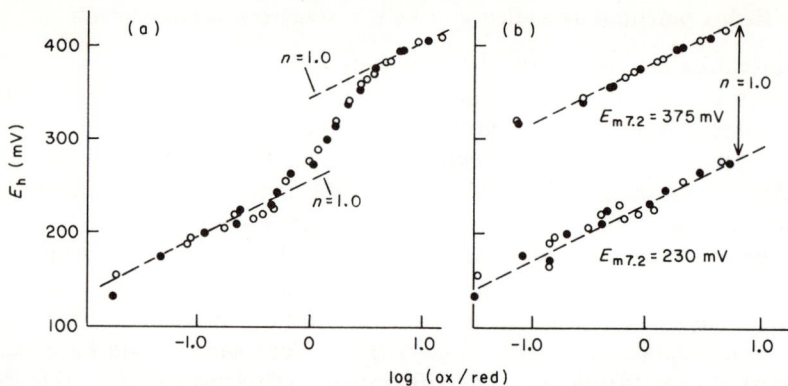

Fig. 5.2 The oxidoreduction potential dependence of the 605 nm minus 630 nm absorbance change of cytochrome oxidase in pigeon-heart mitochondria. (a) The absorbance change is treated as a single component. (b) The sigmoidal curve in (a) is resolved into two $n = 1$ components: O—O, reductive titrations; ●—●, oxidative titrations. From Wilson *et al.* (1972*b*) with permission.

data are obtained in titrations of the 445 nm or Soret band (Wilson and Dutton, 1970).

As concluded in Chapter 4, both redox transitions represent a mixture of both cytochromes *a* and a_3, each exhibiting a high and a low potential form, suggesting that there may be an extensive negative redox interaction between the two haems.

In 1974 Nicholls (1974*b*; Nicholls and Petersen, 1974) proposed a "neoclassical" model in which the classically defined properties of cytochromes *a* and a_3 are retained (Keilin and Hartree, 1939; Vanneste, 1966; Lemberg, 1969; see Chapter 4), and the results from potentiometric titrations (Fig. 5.2) interpreted in terms of a strong haem/haem interaction. The model implies that reduction of either haem decreases the affinity for electrons of the companion haem, and that both haems have originally a roughly equal probability of becoming reduced. This leads to roughly equal E_m values for both haems in the low and high potential redox transitions. Thus, at equilibrium, the half-reduced enzyme will be a mixture of species with either haem *a* or a_3 reduced and the companion haem oxidized (see also Petersen and Andréasson, 1976; Wikström *et al.*, 1976; Babcock *et al.*, 1978).

Malmström (1973) had pointed out earlier that the possibility of redox interactions had been ignored in the original interpretations of the potentiometric titrations, and that such interactions could equally well explain the data as the postulated interactions with respect to spectroscopic properties (Wilson *et al.*, 1972*b*). Wikström *et al.* (1976) showed that the

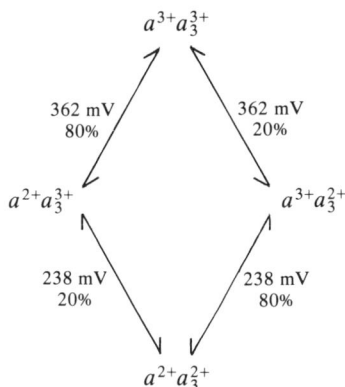

Fig. 5.3 Redox interaction ("neoclassical") model of cytochrome oxidase. Numbers shown represent midpoint potentials at pH 7 and approximate spectral contributions to the absorbance at 605 nm minus that at 630 nm (cf. Wikström *et al.*, 1976).

"neoclassical" model is able to explain these data quantitatively, and that corresponding results in the presence of inhibitory ligands to cytochrome a_3 were in quantitative accordance with known ligand dissociation constants (see also below). Today, a definite choice against large spectroscopic interactions and for negative redox interactions can definitely be made (Wikström *et al.*, 1976; Wikström, 1981*b*; Chapter 4).

Accordingly, we will in the following adopt the "neoclassical" model in an attempt at interpreting redox titration data. It must be stressed that a choice of model is essential, because otherwise there is no single interpretation of the data. Figure 5.3 shows a basic scheme of the neoclassical model as used here (cf. Wikström *et al.*, 1976).

We wish to stress, however, that by choosing this model we do not want to imply that it has been proven correct. In fact, it seems possible that it may have to be revised with an extension of the hitherto proposed interactions between haems a and a_3, to intermonomeric haem/haem interactions in a dimer (Chapter 7). However, the latter model has not been fully developed at this time, and in any case it may be instructive to analyse how far the "neoclassical" model may lead us.

D. The redox properties of Cu_B

Cu_B is not generally detectable by EPR spectroscopy (Chapter 4), nor has it any unambiguously defined band in the visible or near infrared regions. However, the E_m of this interesting centre has been determined by indirect methods, the results of which are in good agreement. Anderson *et al.* (1976) applied a computer fit to reductive titrations with the isolated enzyme (in some experiments cytochrome *c* was also present), and arrived at a value of 340 mV for the E_m. This result is unaffected by the choice of model. Lindsay *et al.* (1975) found that the binding of CO to haem a_3 in intact mitochondria requires that Cu_B is reduced simultaneously. This is a most important finding as it points directly at a close functional co-operativity between the two metals in the a_3/Cu_B centre. Lindsay's conclusion was made on the basis of the finding that the haem a_3–CO complex titrated as an apparent two-electron acceptor in the presence of CO, and that the E_m of the CO compound was raised by 30 mV for each tenfold rise in the CO concentration. The results were tested by simulations using a model in which haem a_3 and Cu_B were assumed to have independent E_m values. Good fits to the data were obtained when the $E_{m,7}$ of Cu_B was set at 340 mV, in agreement with the independent approach of Anderson *et al.* (1976). Lindsay *et al.* (1975) also found that the E_m was unaffected by pH and by the energy state of the mitochondria.

E. The basic models for interpretation of EPR spectroscopy

The $g = 3$ signal of the fully oxidized enzyme can unambiguously be assigned to ferricytochrome *a* (Chapter 4). The dependence of this signal, and the high spin $g = 6$ signals, on redox potential (Fig. 5.1) has been interpreted in two different basic ways.

Either the disappearance of the $g = 3$ signal on decreasing E_h is simply due to reduction of cytochrome *a*, or it may be the result of a low spin to high spin transition in ferric *a*, which also explains the appearance of the $g = 6$ signals. In the former model appearance of $g = 6$ signals must be due to reduction of Cu_B, rendering the high spin ferric a_3 detectable by EPR (cf. Chapter 4). In the latter model, the low spin to high spin transition may be due to haem/haem interaction subsequent to reduction of haem a_3. These two basic models are summarized in Table 5.1.

Table 5.1 shows that the two interpretations lead to opposite assignments of the low potential and high potential transitions in terms of haems *a* and a_3. Both models are associated with conceptual and experimental difficulties. Model A (Table 5.1; cf. Palmer *et al.*, 1976) is beset with the problem that haem a_3 would be expected *a priori* to be the haem with higher E_m due to its reaction with O_2. A more serious argument is the explanation for the appearance of $g = 6$ signals on reduction. The E_m of Cu_B is independent of pH (Lindsay *et al.*, 1975) while the E_m for the appearing $g = 6$ signals is strongly pH-dependent (Fig. 5.1). In model B the $g = 6$ signals are interpreted as arising from haem *a* alone. Yet, there

Table 5.1 Basic interpretation of EPR data from potentiometric titrations.

Model	Midpoint potential		Source of $g = 6$	Explanation of		
	High	Low		disappearing $g = 3$ signal	appearing $g = 6$ signal	disappearing $g = 6$ signal
A	Cytochrome a	Cytochrome a_3	Cytochrome a_3^{3+}	Reduction of cytochrome a	Reduction of Cu_B	Reduction of cytochrome a_3
B	Cytochrome a_3	Cytochrome a	Cytochrome a^{3+}	Spin state transition in cytochrome a caused by reduction of cytochrome a_3		Reduction of cytochrome a

are many instances where $g = 6$ signals in cytochrome oxidase can be attributed unambiguously and uniquely to ferric haem a_3 (Chapter 4).

It is interesting that adoption of the "neoclassical" hypothesis removes most of the difficulties inherent in the primary models in Table 5.1 (see below).

III. Analysis of EPR data according to the neoclassical model

In the cytochrome oxidase monomer, containing four redox centres, there are 16 possible redox state combinations. Of these some appear exceedingly unlikely to be present in significant amounts at equilibrium. This is particularly so for combinations in which Cu_A would be reduced and Cu_B oxidized simultaneously, due to the comparatively well defined E_m values of these centres. The other 11 combinations are listed and numbered by a binary system in Table 5.2. This table also lists our tentative assignment of the observed EPR signals, the basis of which is discussed below.

Table 5.2   Redox state combinations of cytochrome oxidase.

State no.[†]	Cytochrome a	Cytochrome a_3	Cu_A	Cu_B	EPR assignment Haem a	Haem a_3
0	ox	ox	ox	ox	$g = 3$	—
1	ox	ox	ox	red	$g = 3$	$g = 6$[‡]
3	ox	ox	red	red	$g = 3$	$g = 6$[‡]
4	ox	red	ox	ox	$g = 6$	—
5	ox	red	ox	red	$g = 6$	—
7	ox	red	red	red	$g = 6$	—
8	red	ox	ox	ox	—	—
9	red	ox	ox	red	—	$g = 6$[‡]
11	red	ox	red	red	—	$g = 6$[‡]
13	red	red	ox	red	—	—
15	red	red	red	red	—	—

Abbreviations: ox, oxidized; red, reduced.

[†]The redox states are numbered as follows. The redox centres are put in the sequence of the table heading (a, a_3, Cu_A, Cu_B). An oxidized centre is indicated with zero, a reduced centre with 1. The result is the binary form (four digits) of the state number.

[‡]At pH 8.5 part or all of this signal may be converted to the low spin signal with $g = 2.6$.

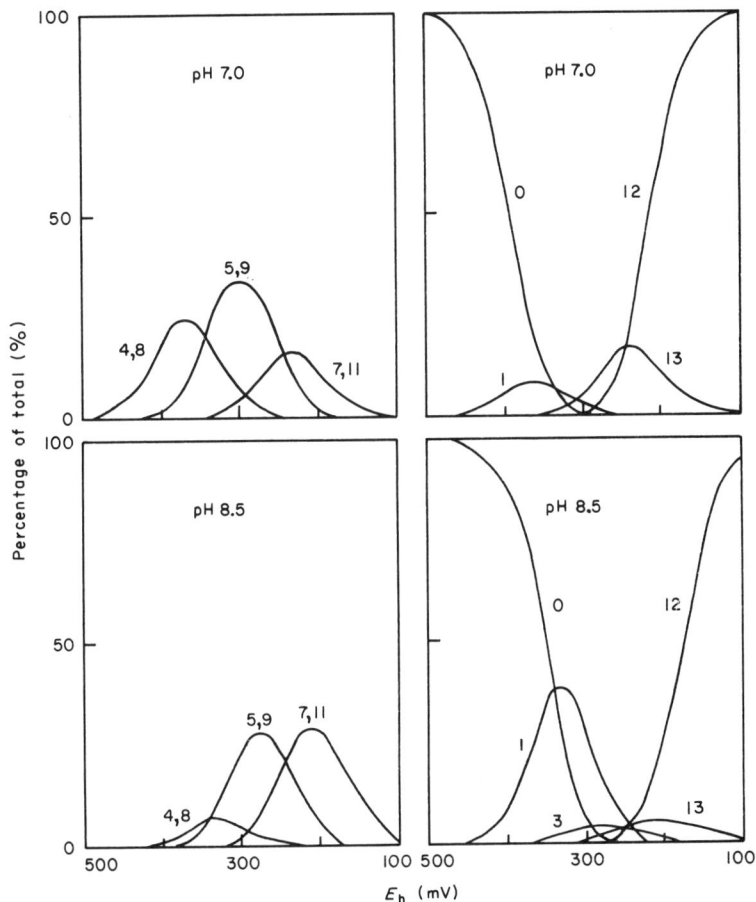

Fig. 5.4 Redox potential dependence of various redox states of the enzyme according to the "neoclassical" hypothesis. Numbers next to the traces refer to the redox states defined in Table 5.2

With the help of the $E_{m,7}$ and $E_{m,8.5}$ values for the haem and copper transitions (Section II), the redox potential dependence of each of the 11 listed states may be simulated. The simulation shown in Fig. 5.4 is based on the neoclassical model (see Fig. 5.3).

A. The assignment of EPR signals

Our analysis may be initiated from the problem that no EPR signal due to Cu_B^{II} is observed under equilibrium conditions, even though reduction of haem a_3 would be expected to break the antiferromagnetic coupling between haem iron and copper. Moreover, the corresponding redox state, i.e. number 4 (see Table 5.2), is expected to contribute significantly at potentials near 360 mV at pH 7 (see Fig. 5.4). The absence of a detectable EPR signal of Cu_B^{II} under these conditions may be due to the high spin state of ferrous a_3 which could result in a symmetry of the coupled haem iron–copper pair low enough to broaden the copper EPR signal beyond detection (Lanne and Vänngård, 1978). Another possibility is antico-operativity between haem a_3 and Cu_B such that reduction of the latter would always precede reduction of the former. Though such an effect may not be unexpected from the proximity of the two metals (Chapter 4), and though it would be consistent with the results of Lindsay *et al.* (1975; cf. Section II.D) as indicated by unpublished simulations (M. Wikström, unpublished), it would introduce differences in the redox state between haems a and a_3 that may be inconsistent with redox titrations monitored by MCD spectroscopy (Babcock *et al.*, 1978), and with the similarity of redox titrations monitored in the α- and Soret bands (Wilson *et al.*, 1972b; Wilson and Dutton, 1970). A third related possibility was proposed by Carithers and Palmer (1981), viz. that the antico-operative effect is, in fact, between haem a and the a_3/Cu_B centre as a whole. Thus the redox state of haem a would not only affect the E_m of haem a_3 but also that of Cu_B. However, a pH-dependent E_m of Cu_B is then expected in contrast to the results of Lindsay *et al.* (1975).

According to the neoclassical model both haems a and a_3 are reduced in the high potential and low potential transitions (see Fig. 5.3). Consequently, the disappearance of the $g = 3$ signal on reduction (Fig. 5.1) is at least in part due to reduction of haem a. The $g = 3$ signal is therefore likely to arise from redox states nos 0, 1 and 3 (Table 5.2). However, states nos 4, 5 and 7 also contain ferric haem a, but cannot contribute to the $g = 3$ signal (compare Figs 5.1 and 5.4), which disappears completely and monophasically with an apparent $E_{m,7}$ of about 390 mV (Fig. 5.1(a)). From this it follows, according to the neoclassical model, that the disappearance of the $g = 3$ signal cannot be *entirely* attributed to reduction of haem a.

Reduction of Cu_B also occurs in this potential range and might thus be responsible for the disappearing $g = 3$ signal. However, reduction of Cu_B is independent of pH (Lindsay *et al.*, 1975), whereas disappearance of $g = 3$ is strongly pH-dependent (Fig. 5.1). Moreover, there is no experimental evidence suggesting an interaction between Cu_B and haem a (but see Carithers and Palmer, 1981, and above). The remaining possibility within the limits of the present model is hence that the $g = 3$ signal disappears, in part, due to reduction of haem a_3, which also occurs in this potential region (Fig. 5.3).

From the above discussion it follows that the disappearance of the $g = 3$ signals in the high-potential region is due to two phenomena, viz. (i) reduction of part of haem a, and (ii) a low spin to high spin transition in the rest of ferric haem a due to reduction of haem a_3 (and haem/haem interaction). According to this, haem a is in a high spin ferric state in redox states nos 4, 5 and 7, and may therefore contribute to the $g = 6$ signals, as indicated in Table 5.2. As discussed below, however (see Section III.B), the notion of a high spin ferric form of haem a is conjectural. On the other hand, the potentiometric titrations monitored by MCD spectroscopy (Babcock *et al.*, 1978; Carithers and Palmer, 1981) provide strong support for the contention that only part of haem a is reduced in the high potential region where the $g = 3$ signal disappears completely. Thus, if the $g = 6$ signals cannot in part be attributed to ferric haem a, then another explanation must be found for why low spin ferric haem a exhibits no EPR signal in the middle range of the redox potential titration.

In redox states nos 1, 3, 9 and 11 the antiferromagnetic coupling between haem a_3 and Cu_B is broken by reduction of the latter. Hence, high spin ferric haem signals ($g = 6$) derived from a_3 are expected in these states (Table 5.2).

At high pH a part of the $g = 6$ signals is replaced by a low spin ferric haem resonance complex ($g = 2.6$; Fig. 5.1(b); Shaw *et al.*, 1978; Lanne *et al.*, 1979). Evidence has been presented indicating that this signal stems from ferric a_3. The strikingly similar properties of this signal with that of metmyoglobin (Gurd *et al.*, 1967), which undergoes a high spin to low spin transition at high pH, further supports this contention. Consequently, the $g = 2.6$ resonance has been assigned to ferric haem a_3 in Table 5.2, for redox states nos 1, 3, 9 and 11 at high pH.

In conclusion, we note that the application of the neoclassical model results in an interpretation that may be considered a compromise between the basic models A and B in Table 5.1.

B. The notion of a high spin ferric form of haem *a*

From the foregoing it is clear that a spin state change in ferric haem a would follow logically from application of the neoclassical hypothesis, and that this is due to haem/haem interaction between cytochromes a and a_3. At present there is, however, little experimental evidence for this notion. The presence of at least two different kinds of high spin ferric haem species with axial and rhombic symmetry is but an indication for the feasibility of this proposal (Wikström *et al.*, 1976). Babcock *et al.* (1978) concluded from a comparison of EPR and MCD data that a sizable fraction of the $g = 6$ signals is due to high spin ferric haem a, but there is unequivocal evidence only for the fact that haem a_3 can contribute to the $g = 6$ signals (Chapter 4).

It may be recalled in this connection that ferric haem a is in the low spin state with a $g = 3$ signal in the half-reduced enzyme where CO or NO is liganded to ferrous haem a_3 (Chapter 4). This contradicts the above conclusion that ferric a becomes high spin on reduction of haem a_3. The neoclassical scheme is therefore forced to add a further condition, viz. that the spin state transition occurs only when haem a_3 is reduced to its high spin state. If so, it follows that photodissociation of CO from the half-reduced enzyme should result in a low spin ($g = 3$) to a high spin ($g = 6$) transition in ferric haem a. Such a phenomenon has indeed been observed (Leigh et al., 1974; Chapter 4). Yet, the uncertainties with respect to interpretation of this effect still leaves the spin state transition in haem a conjectural.

C. Simulation

Using the assignments listed in Table 5.2 and the dependence of the different redox states of the enzyme upon E_h (Fig. 5.4), we can now construct the dependence of the EPR signals on E_h to see how well the hypothesis fits the experimental facts, viz. to the titrations shown in Fig. 5.1. In doing this, we limit ourselves to the high spin and low spin haem signals, discounting the $g = 2$ signal of Cu_A for clarity, and since it is not associated with ambiguity.

The simulations are shown in Fig. 5.5, for pH 7 and pH 8.5. When they are compared with the data in Fig. 5.1 it should be noted that the latter were reported as signal amplitudes in arbitrary units, while Fig. 5.5 is based on actual concentrations, or probabilities, of redox states. Note also that we have, mainly for illustrative purposes, plotted ferric cytochromes a and a_3 separately in their presumptive EPR-detectable states (Fig. 5.5(a)), and further assumed that all EPR-detectable ferric haem a_3 is in the low spin $g = 2.6$ state at high pH (Fig. 5.5(b)).

With these provisos in mind, Fig. 5.5 simulates the experimental data with a fair degree of accuracy, at least on a qualitative basis. Some interesting details may be noted in the comparison between Figs 5.1 and 5.5.

The pH dependence of the transitions is quite nicely simulated. At high pH the rise in $g = 2.6$ signal upon reduction occurs at a higher E_h than the corresponding rise in $g = 6$, which is apparent from both figures. If a sizable fraction of haem a_3 is indeed in the low spin ($g = 2.6$) state at high pH, this would have the effect of "narrowing" the potential region in which the $g = 6$ signals are observed, as compared with the situation at pH 7 (Fig. 5.5). This effect is indeed observed experimentally (Fig. 5.1).

With respect to quantitation, Wilson et al. (1976) reported that about

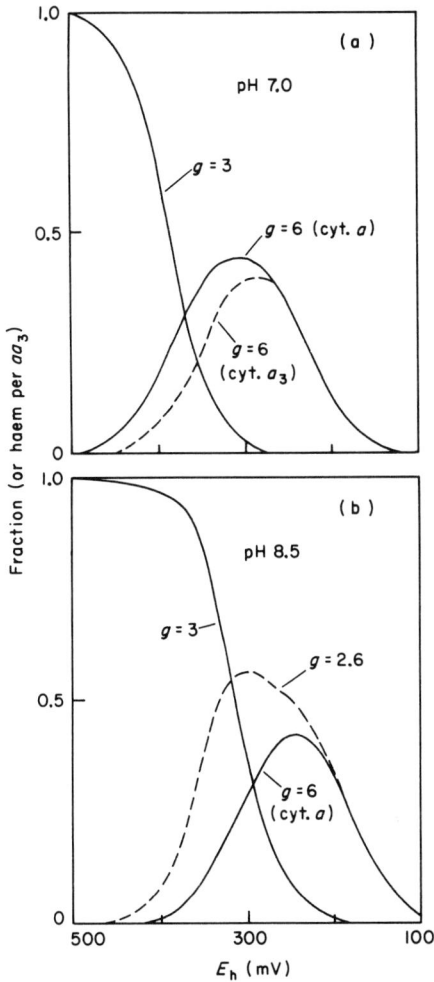

Fig. 5.5 Simulation of the oxidoreduction potential dependence of EPR signals. The simulations at pH 7 (a) and 8.5 (b) are based on the assignments in Table 5.2. At pH 8.5 $g = 6$ signals have been assigned to ferric haem a, and $g = 2.6$ signals to ferric haem a_3.

one haem per aa_3 is detectable in the high spin ferric form by EPR in the half-reduced enzyme. Unfortunately, the EPR signals cannot be very accurately integrated in submitochondrial particles to yield true concentrations. In any case, Fig. 5.5(a) indicates a maximal extent of combined $g = 6$ resonances from haems a and a_3 of about 0.83 haems per aa_3, in good agreement with the experimental estimate.

In the isolated enzyme, however, much less $g = 6$ component is usually observed, usually of the order of 0.3 haems per aa_3 (e.g. Babcock *et al.*, 1978; Hartzell and Beinert, 1976). However, Lanne *et al.* (1979) have more recently observed a maximum value of 0.6 haems per aa_3 as $g = 6$ signal at low pH, and a maximum value of 0.5 haems per aa_3 as combined $g = 6$ and $g = 2.6$ signals at pH 8.3 (with preponderance of the low spin component). Wilson *et al.* (1976) specifically pointed out that the $g = 6$ signals are generally of lower intensity in most isolated cytochrome oxidase preparations as compared to intact mitochondrial membranes. In relation to this, Lanne *et al.* (1979) made the important observation that incorporation of the isolated enzyme in phospholipid membranes had a significant enhancing effect on the $g = 6$ signals. They thus observed up to 0.7 haems per aa_3 as $g = 6$ resonance in such preparations, which comes very close to the "mitochondrial" values.

The fact that the isolated enzyme in detergent solution shows a large fraction of the $g = 2.6$ component also at neutral ambient pH in contrast to the case in the membranous preparations, might furnish an explanation for the relative minute $g = 6$ signals in the former case. This phenomenon, like the low affinity of the oxidized isolated enzyme for cyanide as compared to mitochondria (Chapter 4), again indicates isolation of the enzyme in an "alkaline state". This could be due to a different local proton activity at the enzyme surface depending on whether the surface is in contact with detergent or more natural phospholipid molecules.

IV. The effect of ligands on the redox properties of the haems

A. General remarks

The effect of ligands to cytochrome a_3 on the distribution of electrons among the four prosthetic groups in the enzyme can be a helpful tool in studies of this system. Extensive data are available from the work of Wilson and collaborators (Wilson *et al.*, 1972b), in which the effects of CO, HN_3 and H^+ were studied. Since these results have mainly been discussed in terms of absent redox interactions but with spectral interactions instead, a thorough quantitative analysis of these data may be called for within the framework of the redox interaction (neoclassical) model. A method to this purpose is that of Clark (1960) as adapted by Krab (1977) for a multi-electron acceptor. In its present form this method can also be applied on cases involving either, or both, electron/electron (as in the neoclassical model) and ligand/ligand interactions.

Due to the availability of experimental data, we deal quantitatively with the effects of CO and HN_3 in this section, whereas the effects of HCN are dealt with qualitatively. In Section V we will then deal with the proton, which is of particular importance in the function of the enzyme. A less stringent treatment of the CO and HN_3 cases has been given previously (Wikström et al., 1976).

B. Method for calculation and simulation

Potentiometric titrations are usually analysed with the help of Nernst plots (E_h plotted against log(ox/red); see Fig. 5.2). The shape of a Nernst plot of a two-electron system is completely determined by three parameters (cf. Krab, 1977), so that at any ligand concentration no more than three model-derived parameters may be determined.

The meaning of "E_m" values obtained directly from such Nernst plots (e.g. Fig. 5.2) is model-dependent. Thus, E_m values obtained directly from Fig. 5.2 are true midpoint potentials only in the case that the two redox components exhibit no redox interactions. Yet, this method of obtaining "E_m" values is certainly the simplest one which may be inferred from experimental data. Moreover, such values are widely quoted in the literature. It may therefore be worthwhile to derive the relationship between such "E_m" values and the true model-independent midpoint potentials.

In the following schemes we define a set of constants that are required to describe fully the effect of a ligand L on the redox titration of a two-electron (two-haem)† system.

A ligand may have two effects. By binding differently to the oxidized and reduced forms of a redox couple, it may shift the effective midpoint redox potential. The ligand binding may also change the extinction coefficient (and other spectroscopic properties). Indeed, both effects could take place simultaneously. Scheme A defines these effects for the haem aa_3 system, and these definitions will be the basis for the following mathematical treatment.

In Scheme B the ligand effects on midpoint potential and extinction coefficients have been incorporated into "effective" values of these parameters (indicated by primes) in the presence of ligand, using the unliganded state as reference.

The relationship between these effective values and the values of the corresponding unliganded state is given by the following equations:

$$E_i' = E_i + \frac{RT}{F}\ln\left(1 + \frac{[L]}{K_i}\right) - \frac{RT}{F}\ln\left(1 + \frac{[L]}{K_0}\right), \qquad (5.1)$$

---

†We treat the system here as if no interactions occurred between haems and coppers. The possibilities of such interactions are treated separately in their proper context. Cytochrome oxidase is thus assumed to be a two-electron system.

$$a^{3+}a_3^{3+} - L \qquad (\alpha_{0L})$$

$$K_0 \searrow L$$

$$a^{3+}a_3^{3+} \qquad (\alpha_0)$$

$E_2 \quad e^-$ $e^- \quad E_1$

$$a^{2+}a_3^{3+} - L \xleftarrow[L]{K_2} a^{2+}a_3^{3+} \qquad\qquad a^{3+}a_3^{2+} \xleftarrow[L]{K_1} a^{3+}a_3^{2+} - L$$

$$(\alpha_{2L}) \qquad\qquad (\alpha_2) \qquad\qquad\qquad (\alpha_1) \qquad\qquad (\alpha_{1L})$$

$$E_3 - E_2 \quad e^- \qquad e^- \quad E_3 - E_1$$

$$a^{2+}a_3^{2+} \qquad (\alpha_3)$$

$$K_3 \searrow L$$

$$a^{2+}a_3^{2+} - L \qquad (\alpha_{3L})$$

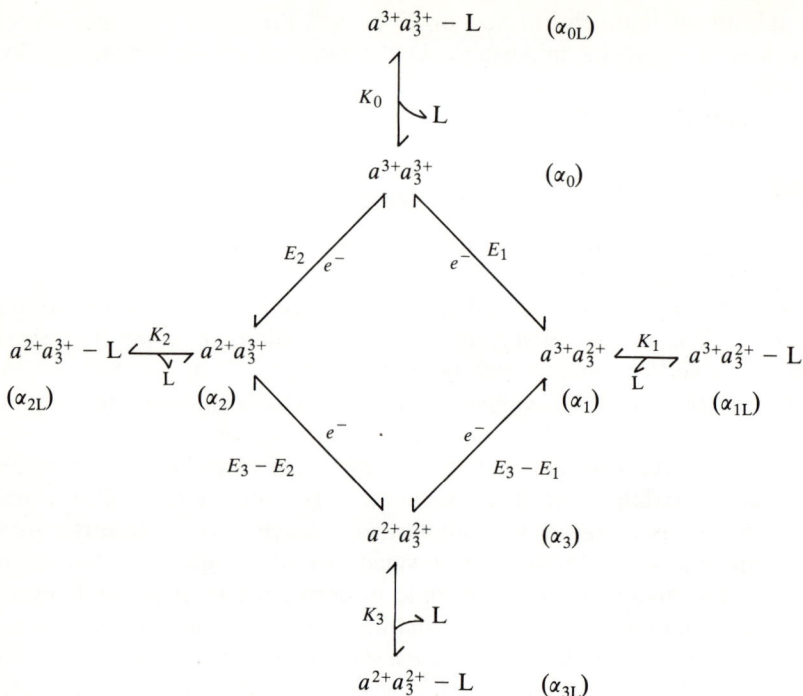

Scheme A   Oxidoreduction midpoint potentials (E), ligand dissociation constants (K) and extinction coefficients (α) for the haem aa_3 system in the presence of a ligand L. Note that E_3 is not (like E_1 and E_2) defined as a midpoint potential but as a sum of two midpoint potentials.

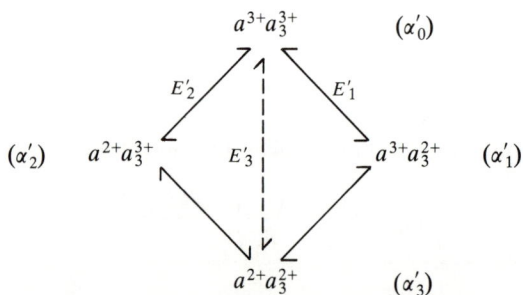

$$a^{3+}a_3^{3+} \qquad (\alpha_0')$$

$E_2' \qquad\qquad E_1'$

$$(\alpha_2') \quad a^{2+}a_3^{3+} \qquad E_3' \qquad a^{3+}a_3^{2+} \quad (\alpha_1')$$

$$a^{2+}a_3^{2+} \qquad (\alpha_3')$$

Scheme B   "Effective" midpoint potentials and extinction coefficients of the haem aa_3 system at a fixed concentration of ligand.

$$\alpha_i' = \left(\alpha_i + \alpha_{iL}\frac{[L]}{K_i}\right)\bigg/\left(1 + \frac{[L]}{K_i}\right). \tag{5.2}$$

Scheme C shows the special case in which no redox or extinction coefficient interactions occur. This is, in principle, the model used by Wilson *et al.* (1972*b*), for which the (apparent) midpoint potentials B are the ones that may be obtained directly from the Nernst plot. In this scheme we have normalized the extinction coefficients β, for convenience.

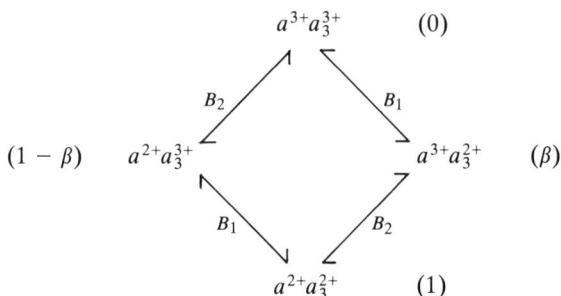

$$a^{3+}a_3^{3+} \qquad (0)$$

$$(1 - \beta) \qquad a^{2+}a_3^{3+} \qquad \qquad a^{3+}a_3^{2+} \qquad (\beta)$$

$$a^{2+}a_3^{2+} \qquad (1)$$

(diagram with B_2 and B_1 labels on the edges)

Scheme C A simple model of non-interacting haems (see text). B_1, B_2 are midpoint potentials. Parameters in parentheses are normalized extinction coefficients for the different species.

The parameters B_1, B_2 and β in Scheme C are now defined as "midpoint potentials" and spectral contribution, respectively, as usually given in the literature whenever these parameters have been directly "read" from Nernst plots.

It is now possible to determine the relationship between these more easily obtainable parameters, and the more general model-independent parameters in Scheme B, at any ligand concentration, provided that a model without interactions fits the data (which is not the case when the Nernst plot indicates positive co-operativity; the expression under the square root (equations (5.3) and (5.4)) then becomes negative):

$$B_{1,2} = \frac{RT}{F}\ln\frac{(e^{x_1} + e^{x_2}) \pm \sqrt{(e^{x_1} + e^{x_2})^2 - 4e^{x_3}}}{2}, \tag{5.3}$$

$$\beta = \frac{1}{2} + \frac{\left(\dfrac{\alpha_1' - \alpha_0'}{\alpha_3' - \alpha_0'} - \dfrac{1}{2}\right)e^{x_1} + \left(\dfrac{\alpha_2' - \alpha_0'}{\alpha_3' - \alpha_0'} - \dfrac{1}{2}\right)e^{x_2}}{\sqrt{(e^{x_1} + e^{x_2})^2 - 4e^{x_3}}}, \tag{5.4}$$

where $x_1 = FE_1'/RT$, $x_2 = FE_2'/RT$ and $x_3 = FE_3'/RT$.

C. Titrations of the 605 nm band in the presence of azide

Azido-cytochrome oxidase readily accepts two electrons when it is titrated with NADH plus phenazine methosulphate (Slater et al., 1965). From this, and the disappearance of the $g = 2$ and $g = 3$ signals (Chapter 4), it may be concluded that haem a and Cu_A become reduced, respectively. Azide therefore seems to stabilize the oxidized forms of haem a_3 and Cu_B.

For these reasons K_1 and K_2 (as defined in Scheme A) may be taken as infinitely large (cf. Wikström et al., 1976). We may also neglect changes in extinction coefficient (Chapter 4), so that $\alpha'_i = \alpha_i$ (as follows from equation (5.2); cf. Schemes A and B), and neglect spectral interactions between haems, so that $\alpha_1 - \alpha_0 = \alpha_3 - \alpha_2$. Five parameters remain to be determined: E_1 (which is equal to E_2 in the neoclassical model), E_3, $(\alpha_1 - \alpha_0)/(\alpha_3 - \alpha_0)$, K_0 and K_2.

E_1 and E_3 are readily determined to be 362 and 600 mV, respectively, from the titrations in the absence of ligands (Wikström et al., 1976). From the known values of B_1, B_2 and β (defined in Scheme C) at any azide concentration (Wilson et al., 1972b), it is possible to calculate K_0, K_2 and $(\alpha_1 - \alpha_0)/(\alpha_3 - \alpha_0)$ by the use of equations (5.1)–(5.4). The average values

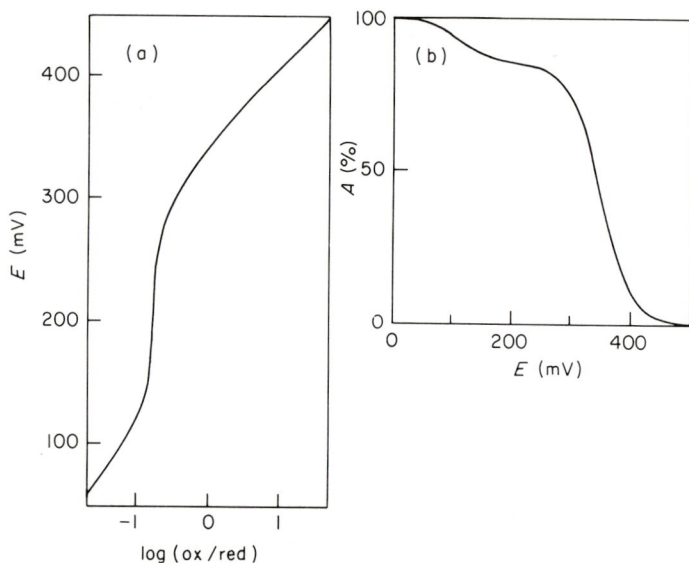

Fig. 5.6 Simulation of potentiometric titration in the presence of 20 mM HN_3. (b) Simulation titration; (a) corresponding Nernst plot. Parameters used: $E_1 = E_2 = 362.3$ mV, $E_3 = 600$ mV, $(\alpha_1 - \alpha_0)/(\alpha_3 - \alpha_0) = 0.142$, $K_0 = 80$ μM, $K_2 = 131$ μM. For definition of parameters, see Scheme A.

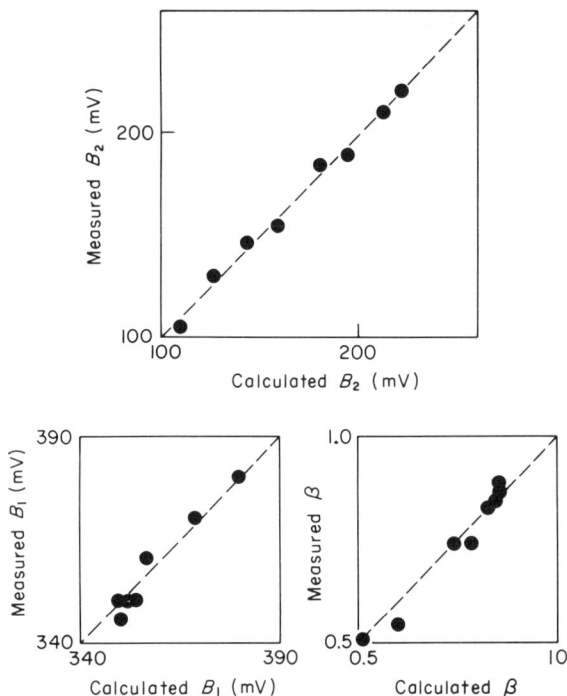

Fig. 5.7 Comparison of experimental and calculated parameters for the simulation of the titration in the presence of azide. Experimental values from Wilson *et al.* (1972*b*). Calculated values were obtained with the parameters used in Fig. 5.6. For definition of designations, see Section IV.B.

obtained this way are $K_0 = 80$ μM, $K_2 = 131$ μM, and $(\alpha_1 - \alpha_0)/(\alpha_3 - \alpha_0) = 0.142$.

The entire potentiometric titration at 605 nm minus 630 nm in the presence of 20 mM azide (see Wilson *et al.*, 1972*b*) may now be simulated using these parameters. The result is shown in Fig. 5.6. Figure 5.7 shows plots of the calculated versus measured values of B_1, B_2, and β at the different azide concentrations used. From both figures it is seen that the fit to the experimental results is quite satisfactory. This simply means that the neoclassical model can quantitatively accommodate the redox titration data in the presence of azide (cf. Wikström *et al.*, 1976). The dissociation constants found for the azide complexes are of the order of magnitude reported previously (Wilson, 1967; Wever *et al.*, 1973). It is interesting that the dissociation constant for the azide complex with ferric haem a_3 depends on the redox state of cytochrome a, as was indeed proposed by Nicholls (1974*b*).

D. Titrations of EPR signals in the presence of azide

The EPR data reported by Wilson *et al.* (1976) of redox titrations in the presence of azide show some interesting features that may provide further information.

In Section III, we concluded that disappearance of the $g = 3$ signal and apparently simultaneous appearance of the $g = 6$ signals in reductive titrations, both at low and high pH in the absence of added ligands (Fig. 5.1), may be the sum of two events: (i) reduction of part of haem a with haem a_3 remaining oxidized; (ii) transition of part of ferric haem a from low spin to high spin in response to reduction of haem a_3. This latter event explains the appearance of part of the $g = 6$ signals; the remaining $g = 6$ signals may be attributed to reduction of Cu_B, which may render ferric haem a_3 detectable by EPR.

At pH 7.2 and in the presence of 60 mM azide, the titration of the EPR signals follows these patterns in principle, but with two essential exceptions (see Wilson *et al.*, 1976). First, no $g = 6$ signals emerge at all, but instead a typical low spin ferric haem signal with $g = 2.9$ appears and disappears much like the behaviour of the $g = 6$ signals in the absence of ligands. However, the disappearance of the $g = 3$ signal and the appearance of the $g = 2.9$ signal are now dislocated from one another (contrast the synchrony of $g = 3$ and $g = 6$ signals in Fig. 5.1), so that the former disappears with an apparent E_m of about 370 mV whereas the latter appears with a much lower apparent E_m, c. 305 mV.

The low spin $g = 2.9$ signal is almost certainly due to ferric haem a_3–HN_3 (Chapter 4), and there is little doubt that the presence of azide prevents significant generation of the species $a^{3+}a_3^{2+}$ during the titration (cf. Wikström *et al.*, 1976). Hence no high spin haem signals are observed from either a or a_3, and the disappearance of the $g = 3$ signal may now be attributed purely to reduction of cytochrome a, in good agreement with the large absorption change at 605 nm during the high potential transition (cf. Fig. 5.6). This is one major difference from the titrations in the absence of ligands.

The data discussed in Chapter 4 indicated that azide causes an optical change consistent with a high spin to low spin transition in ferric haem a_3, when added at high concentrations to the fully oxidized enzyme. Yet, no $g = 2.9$ signal appears under such conditions, but only after partial reduction of the enzyme. By analogy with the case of cyanide (see Babcock *et al.*, 1976), this suggests that azide may react with the oxidized enzyme to form an antiferromagnetically coupled metal pair of the type Fe^{III} ($S = \frac{1}{2}$)–HN_3–Cu_B^{II} ($S = \frac{1}{2}$) with total spin $S = 0$ in the ground state. This would make the complex undetectable by EPR at low temperatures.

The proposal may be tested by MCD spectroscopy and magnetic suscepti-bility measurements.

The finding that cytochrome oxidase readily accepts only two electrons per aa_3 in the presence of azide (Slater et al., 1965; see above) indicates stabilization of *both* ferric haem a_3 and Cu^{II} by azide, with concommittant lowering of the E_m of both redox couples. This would be consistent with a bridged complex. If the E_m of Cu_B is indeed lowered to about 300 mV by this mechanism, it would explain why the $g = 2.9$ signal appears in the reductive titrations with half-reduction potential of this magnitude (see above). It is significant that this value is unaffected by pH (Wilson et al., 1976) in accordance with the earlier findings that the E_m of Cu_B is inde-pendent of pH (Section II). From this discussion it also follows that the strong pH dependence of the $g = 3$ signal in the presence of azide (Wilson et al., 1976) must then under these conditions be entirely attributed to a strong pH dependence of the E_m of cytochrome a (cf. Section V).

E. Carbon monoxide

Unlike azide, CO binds to ferrous haem a_3. Hence $K_0 = K_2 = \infty$. The relevant K_D values ($K_3 = 0.33$ µM; $K_1 = 0.59$ µM) are obtained from the work of Greenwood et al. (1974). Again it seems that the redox state of cytochrome a exerts an influence on the affinity of haem a_3 for ligands (cf. azide; Section IV.C).

The simulation of the potentiometric titration in the presence of 1 mM CO, and the corresponding Nernst plot, are shown in Fig. 5.8. The exact concentration of CO is not critical, provided that it is much higher than the dissociation constants. The only difference between Fig. 5.8 and the experimental results (cf. Wilson et al., 1972b) is the deviation of the Nernst plot from a straight line in the former case. This is because the true 100% oxidation level of the enzyme has been used to calculate log(ox/red) in the simulation, whereas this level is never reached experimentally due to the very high E_m of the a_3–CO complex. If the dashed line in Fig. 5.8(b) is taken as the 100% oxidized level (where a_3 is still reduced and liganded to CO), the simulation conforms quite well to the experimental data (see dashed line in Fig. 5.8(a)). The apparent E_m of haem a is 253 mV from the simulation, which is in excellent agreement with the values obtained experimentally (Tzagoloff and Wharton, 1965; Wilson et al., 1972b).

F. Cyanide

Redox titrations of isolated or membranous oxidase in the presence of cyanide have been performed by several authors (Minnaert, 1965; Hinkle and Mitchell, 1970; Wilson and Leigh, 1974; Artzatbanov et al., 1978). At

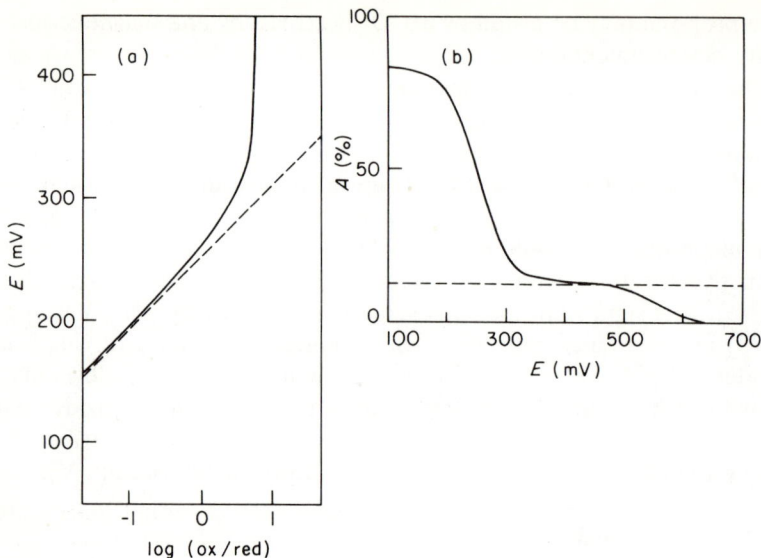

Fig. 5.8 Simulation of potentiometric titrations in the presence of CO. (b) Simulated titration; (a) corresponding Nernst plot. Parameters used: $E_1 = E_2 = 362$ mV, $E_3 = 600$ mV, $K_1 = 0.59$ µM, $K_3 = 0.33$ µM, $\alpha_{1L} = 0.13$, $\alpha_{3L} = 0.83$. The source of dissociation constants and extinction coefficients is Greenwood *et al.* (1974). For an explanation of the designations, see Section IV.B and Scheme A. [CO] = 0.93 mM.

least when carried out aerobically, cyanide is expected to "clamp" haem a_3 effectively in the oxidized state throughout the titration. Under such conditions the $E_{m,7}$ of cytochrome a has been found to be between 270 and 310 mV. However, all groups cited above (except Wilson and Leigh, 1974) report that the titration of haem a is anomalous in that Nernst plots yield an apparent $n = 0.5$ ("half-electron carrier"). This phenomenon has remained unexplained to date (cf. Slater *et al.*, 1965, for some possibilities).

   If two $n = 1$ (one-electron) carriers that absorb at the same or similar wavelengths have E_m values about 60–100 mV apart from one another, a titration of the "mixed" absorption band with treatment of the absorption change as if due to a single component can yield apparent $n = 0.5$ behaviour over a large range of E_h (up to about 200 mV). The anomaly may thus reflect an inhomogeneity of the redox behaviour of cytochrome a, which may be due to at least two different reasons. One is an antico-operativity between haem a and Cu_A (for which there is no evidence), and the other an antico-operativity between two cytochromes a (cf. Chapter 7),

However, we have no explanation for the fact that the phenomenon is apparently observed only with cyanide, and not, for example, in the presence of azide or in the absence of ligands.

V. pH dependence of the midpoint potentials of the haems

A. The involvement of protons

Hydrogen ions are of fundamental importance in the cytochrome oxidase reaction. In the first place H^+ is a substrate required in the generation of water following electron transfer to O_2. Secondly, cytochrome oxidase conserves a large fraction of the free energy change by catalysing proton translocation across the mitochondrial membrane (Wikström and Krab, 1979; Chapter 7).

The E_m values of the haems, but not those of the coppers, are pH dependent. This is highly significant in relation to the proton translocation mechanism (Chapter 7). Here we only attempt to give a quantitative description of the pH dependence for the enzyme in detergent solution. The proton is thus considered as a ligand in precisely the same way as above for azide and carbon monoxide.

The earliest available data (Wilson et al., 1972b) are titrations of cytochrome oxidase in rat liver or pigeon heart mitochondria in the presence of uncoupling agents, or otherwise under conditions where no difference in proton activity would be expected across the mitochondrial membrane. Here we nevertheless use the more recent and more extensive data of Van Gelder et al. (1977), which were obtained with the isolated beef heart enzyme. However, these data are largely similar to those obtained with the enzyme in situ.

There is no extensive effect of pH on the absorption at 605 nm. Hence we may put $\alpha'_i = \alpha_i$ in our calculations. We maintain the values obtained for the (fractional) extinction coefficients α in the titrations with azide or CO (i.e. 86% contribution of haem a at 605 nm; cf. Chapter 4). However, this time we must take into account possible liganding of all four redox states of the oxidase haems with H^+, viz. $a^{3+}a_3^{3+}$, $a^{2+}a_3^{3+}$, $a^{3+}a_3^{2+}$ and $a^{2+}a_3^{2+}$.

B. Simulation

A first glance at the pH dependence of the "midpoint potentials" B_1 and B_2 (cf. Van Gelder et al., 1977; Wilson et al., 1972b; Fig. 5.9) reveals that it would not be possible to simulate this behaviour with one acidic group only, having a maximum of four K_D values: K_0, K_1, K_2 and K_3. A second acid group must be introduced, and for simplicity it is assumed that the two

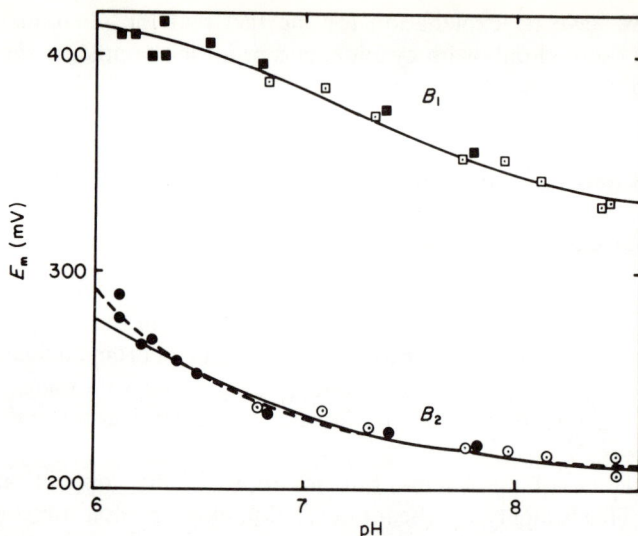

Fig. 5.9   pH dependence of the E_m of the two redox transitions of the cytochrome oxidase haems. Experimental points from the data of Van Gelder *et al.* (1977). Open symbols, in the presence of cholate; solid symbols, in the presence of Triton X-100. The lines are calculated with parameters of the first column in Table 5.3. The dashed line is calculated with parameters of the third column (three acid/base groups; Table 5.3).

groups are non-interacting. The equations for E_i' then take the form

$$E_i' = E_i + \frac{RT}{F} \sum_{j=1}^{2} \{\ln(1 + 10^{pK_{ij}-pH}) - \ln(1 + 10^{pK_{0j}-pH})\}, \quad (5.5)$$

in which K_{ij} is the dissociation constant of the jth acid group in redox state i (cf. equation (5.1); Section IV.B). When necessary, this equation can easily be extended to include mutual interactions between acidic groups (see Krab, 1977).

The contribution of high potential and low potential redox components to the absorption at 605 nm is about 50% and nearly independent of pH (Wilson *et al.*, 1972b; Artzatbanov *et al.*, 1978). When it is assumed, as before, that $\alpha_0 - \alpha_1 = \alpha_3 - \alpha_2$ (see Section IV.C), it may be shown that this means that $E_1 = E_2'$ at all pH values involved. Hence, not only $E_1 = E_2$ (which is the equivalent of the neoclassical axiom at $[H^+] = 0$), but it also follows that the pK values of the groups involved must be very similar in the redox states corresponding to the half-reduced enzyme, viz. $a^{2+}a_3^{3+}$ and $a^{3+}a_3^{2+}$. We assume here, therefore, that they are indeed similar, which enables us to reduce the number of fit parameters to eight (i.e. two times three pK values plus E_1 and E_2).

In Fig. 5.9 a fit to the data of Van Gelder et al. is presented with the set of parameters listed in Table 5.3. Apart from the relatively well defined pK values in the pH range actually covered by the experiment, a low pK was required to account for the steep rise of the lower midpoint potential (B_2) at low pH. Even then, the fit is not very satisfactory in this region. The dashed line in Fig. 5.9 gives the result of an attempt to improve the fit by introduction of a third acidic group (cf. Table 5.3). Although the resulting fit is then quite good, data of other groups (Wilson et al., 1972b; Artzatbanov et al., 1978) do not indicate a similar increase of the E_m at low pH. From Fig. 5.9 (see curve of B_1) we also get the impression that there may be a substantial variation in the experimental data in the low pH region. This suggests that it may be premature to add a third acid group.

A similar fit (not shown) is also obtained for the data of Artzatbanov et al. (1978), the parameters of which are also given in Table 5.3.

Finally, we should emphasize that the fits shown here are not "best fits", but have been constructed more loosely. This seems justifiable due to the rather extensive scattering of the experimental data. It follows, of course, that the obtained apparent pK values (Table 5.3) are only approximate, giving the general trend of the system's behaviour.

C. Significance of the pK values found

We may conclude that, outgoing from the neoclassical model, the pH dependence of the haem's redox potentials may be satisfactorily simulated by invoking only two different redox-linked acidic groups. However, it must be stressed that we cannot exclude the possibility that there are many more such groups. If so, the listed pK values are "composite" and give the impression of the summated function of all redox-linked acid groups present.

Table 5.3 indicates a priori that none of the acid residues is specifically associated with either haem group since no pK change seems to follow the redox pattern of any single haem. This leads to an interesting conclusion. First, there are obviously at least two acidic residues in the enzyme, the pK values of which are linked to the haems' redox state, but more important, the changes in pK seem to occur in response to a change in redox state of the two-haem system as a whole. It appears that one acidic group is sensitive to the transition between fully oxidized and half-reduced enzyme, irrespective of which haem becomes reduced, whereas the other acid group senses the transition between the half-reduced and fully reduced states. If this is true, then the phenomenon is clearly relevant with respect to the

Table 5.3 pK values of redox-linked acid/base groups used to fit the data in Fig. 5.9. Parameter denotations are defined in Section IV.B.

Redox state	Two acid/base groups (Fig. 5.9; continuous line)		Two acid/base groups (data of Artzatbanov et al.)		Three acid/base groups (Fig. 5.9; dashed line)		
	Group 1	Group 2	Group 1	Group 2	Group 1	Group 2	Group 3
$a^{3+}a_3^{3+}$	4.0	6.4	6.0	6.8	4.5	4.5	6.4
$a^{3+}a_3^{2+}$	4.0	8.0	6.0	7.9	4.5	4.5	8.0
$a^{2+}a_3^{3+}$	4.0	8.0	6.0	7.8	4.5	4.5	8.0
$a^{2+}a_3^{2+}$	7.0	8.0	6.9	8.0	6.5	6.5	8.2
Other parameters							
$E_1 = E_2$	310 mV		284 mV			310 mV	
E_3	533 mV		477 mV			535 mV	
$(\alpha_1 - \alpha_0)/(\alpha_3 - \alpha_0)$	0.14		0.14			0.14	

strong interactions between the two haems proposed in the neoclassical model.

This picture is supported by the finding that the redox transition in haem a is highly pH dependent when haem a_3 is "clamped" in the oxidized state by HCN (Artzatbanov *et al.*, 1978) or by azide (Wilson *et al.*, 1978; Section IV.D), but shows no or only very slight pH dependence when haem a_3 is "clamped" in the reduced state by CO (Hinkle and Mitchell, 1970; K. Krab and M. Wikström, unpublished). Analogously, it is the high potential redox transition in the haem aa_3 system that shows a strong pH dependence, whereas the low potential transition is much less pH dependent at pH values around 7 (Wilson *et al.*, 1972b; see Fig. 5.9).

Due to the apparent mutual coupling of the two haems to common acidic residues, the redox states of the two haems would of necessity be interdependent. We may therefore pose the question whether this coupling could be the sole explanation for the "splitting" of the E_m values of each haem in two, depending on the redox state of the companion haem. That this would not be the case follows in principle from the values of E_1, E_2 and E_3 (Table 5.3). But we cannot exclude, of course, that there may be further pH effects in a range outside that covered by Fig. 5.9, such that in the final analysis $E_3 = 2E_1$.

VI. Conclusions

From the foregoing chapter we may conclude that the neoclassical model provides a framework within which the equilibrium behaviour of cytochrome oxidase may be successfully described. It is thus possible to describe the origins and properties of different EPR signals of the redox centres in a quantitative way (Sections II and III), and to describe the effect of extraneous inhibitory ligands, especially CO and HN_3, in a way that is in quantitative agreement with the experimental data (Section IV). It is therefore clear that qualitative criticism of this model based on the redox behaviour of the system in the presence of ligands (see Erecińska and Wilson, 1978) is entirely unfounded.

An interesting possible extension of the concept of haem/haem interaction has been the finding that the enzyme contains at least two redox-coupled protonable groups, the pK values of which may depend on the redox transitions of the haem system as a whole rather than on the individual cytochromes (Section V). This suggests a possible explanation for the observed redox antico-operativity and might become the harbinger of our future understanding of this phenomenon.

We should stress, however, that the above points have been arrived at under the assumption that haem/haem interactions may only take place between haems a and a_3 in monomeric cytochrome oxidase. Since there are now growing indications that a fundamental property of the oxidase may be to function as the $(aa_3)_2$ dimer (Chapter 7), it is clear that the data analysed in this chapter may have to undergo a profound re-examination in which intermonomeric interactions are also allowed.

6

Kinetics and catalytic mechanism

I. Introduction

The final word on an enzyme's catalytic mechanism must come from kinetics, even though any proposed mechanism must also be consistent with the available structural and thermodynamic information.

In this chapter we will discuss the main kinetic studies with cytochrome oxidase, which may be divided grossly into studies on the reaction with O_2 and the reaction with ferrocytochrome c. One of our goals is to demonstrate that temporally and technically different studies may be brought together to provide a unique picture. Part of the information on electron transfer is closely related to energy transduction, and is therefore discussed in more detail in Chapter 7.

II. On the reduction of dioxygen to water

It may first be valuable to discuss briefly the chemistry of O_2 reduction. The properties of dioxygen and its reduction were treated in detail in a lucid article by George (1965) on "The fitness of oxygen". Some points may warrant examination here in the light of information obtained more recently.

The four elementary one-electron steps of O_2 reduction through peroxide are summarized in Table 6.1, along with their $E_{m,7}$ values. Clearly, these thermodynamic parameters must be altered by binding of the oxygen species to the a_3/Cu_B centre. Yet, the table emphasizes some problems that may be relevant in considerations of the catalytic mechanism.

The one-electron pathway involves the radicals $O_2H^.$ and $HO^.$, of which particularly the latter is highly reactive. Thus a minimum requirement is that these species, if formed, remain tightly bound to the "active site". However, their reactive chemistry might well have led to evolution of a catalytic mechanism such as to avoid them.

The second problem relates to the strikingly low $E_{m,7}$ for the transfer of the first electron to O_2 (Table 6.1). Since the O_2-reducing centre operates at redox potentials well above 0.5 V (cf. Erecińska and Wilson, 1978), this first step would be exceedingly unlikely and therefore very slow. Even if

117

Table 6.1   The reduction of dioxygen to water via peroxide at pH 7.

Electrode reaction	$E_{m,7}$ (V)	References
1. $O_2 + e^- \longrightarrow O_2^{\cdot -}$	-0.33	Wood, 1974
2. $O_2^{\cdot -} + e^- + 2H^+ \longrightarrow H_2O_2$	$+0.94$	Wood, 1974
3. $H_2O_2 + e^- + H^+ \longrightarrow HO^{\cdot} + H_2O$	$+0.31$	†
4. $HO^{\cdot} + e^- + H^+ \longrightarrow H_2O$	$+2.33$	George, 1965
5. $\frac{1}{2}O_2 + e^- + H^+ \longrightarrow \frac{1}{2}H_2O_2$	$+0.305$	Wood, 1974
6. $\frac{1}{2}H_2O_2 + e^- + H^+ \longrightarrow H_2O$	$+1.32$	†
7. $\frac{1}{4}O_2 + e^- + H^+ \longrightarrow \frac{1}{2}H_2O$	$+0.81$	George, 1965

The values given refer to an aqueous solution at pH 7. Standard state of dioxygen refers to a fugacity of 1 atm, for other species 1 molal solutions. Temperature 25 °C. Data from George (1965) as amended by Wood (1974).

†These values are calculated on the basis of the values given by Wood for reactions 1, 2 and 5, and the values given by George for reactions 4 and 7.

binding of the superoxide radical to the enzyme were much stronger than binding of O_2, causing an enhancement of the E_m, the differential binding affinity can hardly be large enough to overcome the tremendous thermodynamic barrier (see George, 1965; Wilson et al., 1977; Erecińska and Wilson, 1978).

As discussed in Chapters 4 and 5, both structural and functional studies have over the recent years strongly indicated that a binuclear haem a_3/Cu$_B$ centre is the site of O_2 reduction. Together with the thermodynamic argument above, this suggests that the reduction of dioxygen to peroxide may take place by a concerted two-electron transfer (Erecińska and Wilson, 1978; Reed and Landrum, 1979). To be effective in overcoming the thermodynamic barrier, this requires that the two electrons are transferred in sufficiently fast succession so that no diffusion of partially reduced intermediates ($O_2H^{\cdot}$) is possible, and with no time for relaxation of the nuclear configuration of this intermediate before arrival of the second electron.

The third one-electron step is also more unfavourable energetically than the fourth (Table 6.1). This is because in step 3 the O—O bond cleavage must be coupled to the reduction. Here the $HO^{\cdot}$ radical is formed, one of the most reactive compounds in chemistry. One is therefore inclined to propose that the reduction of peroxide to water might also occur by an effective two-electron transfer.

A tentative mechanism will be proposed below which conforms to these requirements. First, however, we should acquaint ourselves with the kinetics of the O_2 reaction, which has been studied by rapid flow and photolysis techniques, and more recently at very low temperatures.

III. Kinetics of the reaction with O_2

At room temperature reduced cytochrome oxidase reacts with O_2 with a second-order velocity constant of about $10^8 \, M^{-1} \, s^{-1}$ (Gibson and Greenwood, 1963, 1965; Chance and Schindler, 1965; Greenwood and Gibson, 1967; Chance and Erecińska, 1971). Figure 6.1(a) shows that the kinetics are biphasic when measured at 605 nm, as they are at 445 nm. Oxidation of Cu_A as measured at 820 nm (see Chapter 4) is also biphasic. At 445 and 605 nm the biphasic kinetic oxidation pattern has been interpreted classically in terms of a simple linear electron transfer sequence, viz. $O_2 \leftarrow a_3 \leftarrow a \leftarrow c$, corresponding to the classical picture of electron transfer in the terminal segment of the respiratory chain (see also Erecińska and Chance, 1972, for measurements at temperatures down to $-30 \, °C$). According to this view the fast phase is the result of oxidation of haem a_3 while the slow phase is due to oxidation of haem a. However, as extensively discussed previously (for a review see Lemberg, 1969), this interpretation is beset with the difficulty that the extent of the fast phase at 605 nm (c. 40% of the total reduced minus oxidized difference spectrum) is much larger than expected from the contribution of haem a_3 at this wavelength. This led to various proposals of different "forms" of haem a_3 with different spectral properties in kinetic and static experiments. However, when the data from potentiometric titrations became available (Wilson et al., 1972b; see Chapters 4 and 5) the problem was assumed to be solved in terms of the new assignment of equal contribution of haem a and a_3 to the 605 nm band (see Erecińska and Chance, 1972). Nicholls (1974b) and Nicholls and Chance (1974) have most clearly emphasized how the problem of the spectroscopic assignment is related to the interpretation of the kinetics of oxidation.

In Chapter 4 we concluded that the original spectral assignments (Keilin and Hartree, 1939) are correct, cytochrome a being the main absorbing species of the reduced minus oxidized band at 605 nm. From this it follows that the interpretation of the kinetics (Fig. 6.1(a)) still presents a real problem. As will be shown below, however, this problem finds a unique solution that may yield new and invaluable information about the catalytic mechanism.

Before proceeding further we wish to emphasize another detail from the pioneering work of Gibson and Greenwood (1965). When the kinetics are measured at 610 nm instead of the more conventional 605 nm (Fig. 6.1(b)) the first rapid absorption decrease is actually followed transiently by an increase in absorption before the slow phase of absorption decrease commences. Also this "anomaly" finds an interesting explanation when related to the more recent low temperature data.

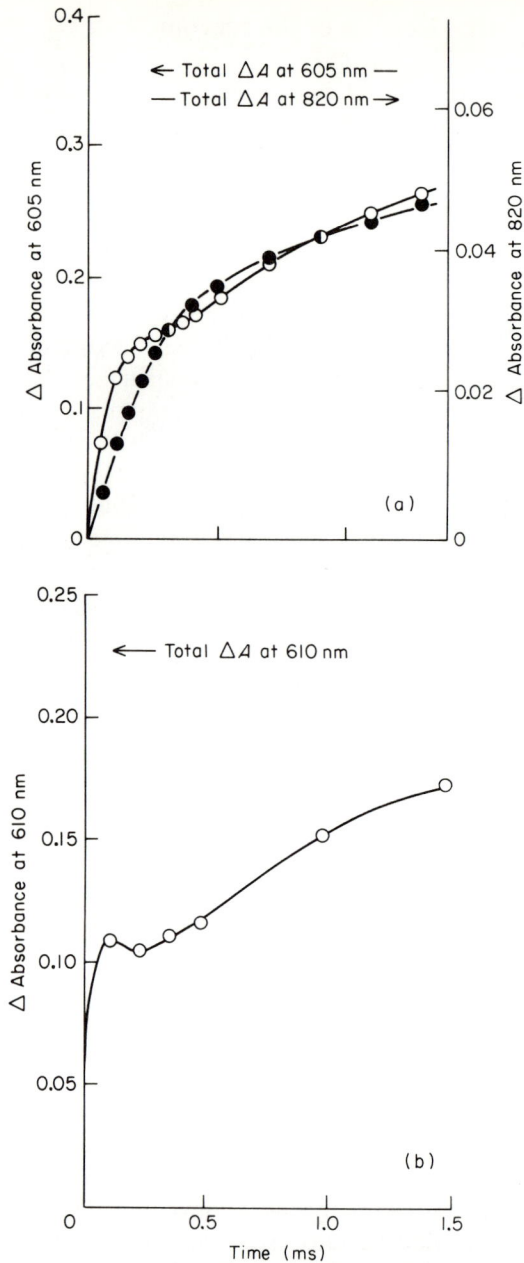

Fig. 6.1 The oxidation of reduced cytochrome oxidase by O_2 at room temperature. In (a) the reaction is followed at 605 (O—O) and 820 nm (●—●). In (b) the reaction is followed at 610 nm. Upward deflection reflects a decrease in absorption. From Gibson and Greenwood (1965) with permission.

IV. Intermediate steps in the reduction of dioxygen

Resolution of the individual catalytic steps of O_2 reduction by cytochrome oxidase has only recently become possible after introduction of the ingenious "triple-trapping" technique by Chance, Saronio and Leigh (1975). In this technique the enzyme, or the mitochondrial suspension, is first saturated with carbon monoxide under reducing conditions to form the ferrous a_3–CO compound, and then cooled in the presence of "antifreeze" (usually ethylene glycol) to about -20 to -30 °C in the liquid state. At this point O_2 may be added to the suspension by stirring or by mixing with O_2-saturated medium. The O_2 will not react with the enzyme due to the very slow dissociation of CO from haem iron in the dark at this temperature. Subsequently, the oxygenated suspension is frozen in an alcoholic dry ice bath (-78 °C) and then placed in a thermostatted cuvette in which temperature equilibrium occurs to a final reaction temperature anywhere between -130 and -60 °C. The reaction with O_2 may now be initiated by an intense flash of light, causing photolysis of the iron–CO bond. At temperatures below about -60 °C the recombination of CO with haem iron is sufficiently slow not to interfere with the oxygen reaction. The reaction is followed spectrophotometrically and may be stopped at any point by rapid freezing in liquid N_2 for further studies of optical or EPR spectra, or by other spectroscopic techniques.

A. The "oxy" species or Compound A

On photolysis of the reduced enzyme, the first event observed spectroscopically is a decrease of the 590 nm band of the a_3–CO species and appearance of a new band of approximately equal magnitude near 612 nm, presumably due to generation of five-co-ordinate haem a_3^{2+} (cf. Fig. 6.2; Chapter 4). This is the classical CO dissociation difference spectrum. No further changes are observed (except for an extremely slow recombination of CO) unless O_2 was added prior to freezing.

If O_2 is present, and at a temperature higher than $c.$ -100 °C, the above is followed by spectral changes that are roughly the inverse of those due to photolysis (Fig. 6.2; see also Clore et al., 1980a). Apart from the requirement for O_2, the compound formed cannot be photolysed and is further converted to other intermediates (see below). Clearly, the effect cannot be due to recombination of CO with haem iron.

From the spectral change (Fig. 6.2) it is evident that the newly formed compound (Compound A) is spectrally similar to the Fe^{II}–CO compound of haem a_3. The same similarity is also apparent in the Soret region. These data are therefore suggestive of the structure

$$Fe^{II}\text{–}O_2Cu_B^{I} \tag{6.1}$$

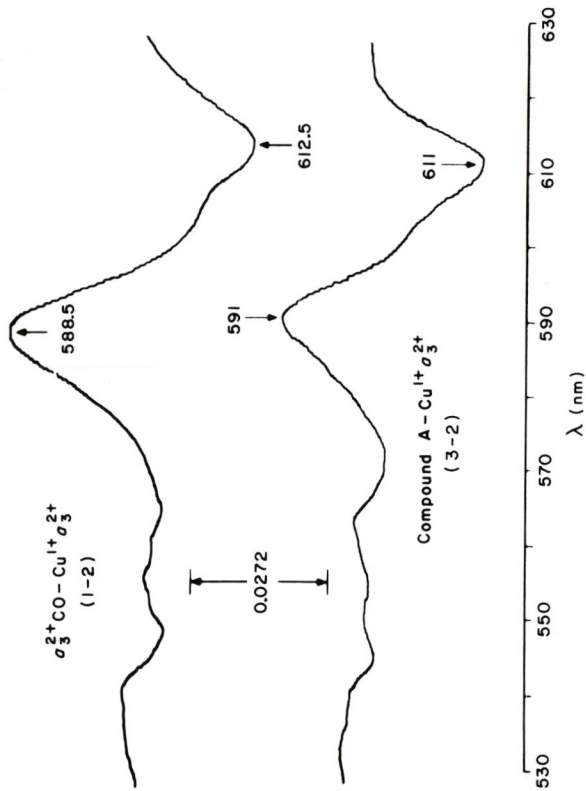

Fig. 6.2 Flash-photolysis of reduced cytochrome oxidase in the presence and absence of O_2. Reaction temperature $-97\,°C$. Dual wavelength traces are shown to the right in the absence (trace I) and presence (trace II) of O_2. The spectra to the left designated 1–2 and 3–2 shows the changes between states numbered correspondingly in traces I and II, respectively. From Chance et al. (1975) with permission.

proposed by Chance *et al.* (1975). This is also consistent with the absence of any evidence for oxidation of any redox centre in the enzyme. Model haem studies by Babcock and Chang (1979) also support this contention, and suggest that O_2 is terminally bound to haem iron rather than "bridged" between iron and copper. This also agrees with the case of the analogous CO compound, which was studied by infrared spectroscopy (Caughey *et al.*, 1976). Clore *et al.* (1980*a*) pointed out the probability of partial charge transfer from iron to bound dioxygen in this species, as has also been inferred in oxyhaemoglobin from the infrared stretching frequency of the O—O bond of the haem-bound dioxygen (Barlow *et al.*, 1973; for a review see Makinen, 1979). All this suggests then that Compound A may indeed be the long-sought-for "oxy" intermediate.

It is interesting that the low affinity of haem a_3 for O_2 at low temperature ($K_D \simeq 0.3$ mM; Chance *et al.*, 1975) agrees well with expectations based on the structure of haem A. The substituent formyl group on the porphyrin ring (Chapter 2) provides strong electron-withdrawing power when conjugated with the ring. This is expected to reduce the affinity of the haem iron for ligands (Babcock and Chang, 1979), but may be necessary to provide a sufficiently high midpoint redox potential. The high apparent affinity of the oxidase for dioxygen at room temperature (see Nicholls and Chance, 1974) may then be the result of very rapid electron transfer to bound dioxygen, before it has time to dissociate from the iron (Chance *et al.*, 1975). This would preclude any significant probability of Compound A "occupancy" during turnover at high temperatures.

Although the identification of Compound A seems reasonably certain, its accumulation at low temperature might be an artefact of the triple-trapping technique. Alben *et al.* (1981) have recently shown by Fourier transform infrared spectroscopy that photolysis of the Fe^{II}–CO at low temperature leads to displacement of the CO from iron to copper. The Cu^I–CO is formed at temperatures below about -73 °C. This would explain nicely why recombination of CO with haem iron is so anomalously slow in cytochrome oxidase, when compared with other haemoproteins (Yonetani *et al.*, 1973). Thus binding of CO to copper might stabilize Compound A artificially. On the other hand, the data clearly demonstrate (Chance *et al.*, 1975) that the reaction proceeds beyond the Compound A stage. This is consistent with the findings by Alben *et al.* (1981), because the Cu^I–CO compound was stable only below -133 °C. Thus, even if Compound A accumulates artefactually, further steps are yet likely to reveal normal catalytic events that are unaffected by the use of CO in the low temperature technique. As shown recently (Wikström, 1981*c*, and see below), this view is strongly supported by data on "reversed electron transfer" in mitochondria.

B. The "peroxy" intermediate and Compound C

Whereas both the half-reduced oxidase (i.e. haem a_3 and Cu_B reduced; Chapter 5) and the fully reduced enzyme form Compound A on reacting with O_2 (Chance et al., 1975; Clore et al., 1980a,b), further steps of the respective reactions have been described as very different indeed. Figure 6.3 shows the difference spectra of end products at $-78\ °C$ for these two cases (relative to the initial carbonmonoxy-enzyme), i.e. so-called Compounds C and B, respectively.

The reaction of the half-reduced or "two-electron" oxidase with O_2 may be expected to be much simpler than the reaction of the fully reduced enzyme, in which electron transfer from haem a and Cu_A may also occur (see below). We therefore consider Compound C first.

Compound C has intriguing spectral properties. Concomitant with the disappearance of the spectral attributes of Compound A, there is a large

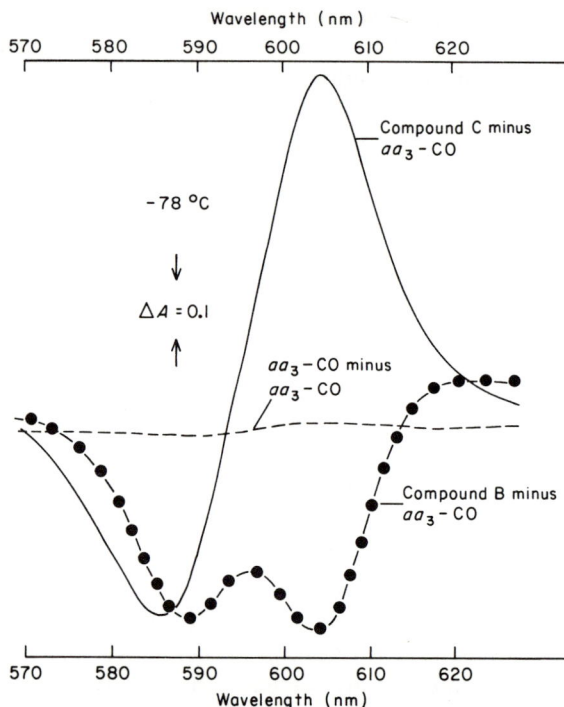

Fig. 6.3 Difference spectra of Compounds B and C relative to the reduced CO state of the fully reduced and half-reduced enzyme, respectively. Reaction temperature $-78\ °C$. (Unpublished experimental results obtained by M. Wikström and K. Krab.)

increase of absorption intensity, peaking in the 605–610 nm region (Fig. 6.3). However, the corresponding Soret band is very weak in comparison to the band of ferrous a_3 in the reduced enzyme (Chance et al., 1979; Denis, 1981). The extinction of the 605 nm band may be estimated to be about 12 mM^{-1} cm^{-1} when "extrapolated" to room temperature (relative to the fully oxidized enzyme; based on the aa_3 content) by comparison with extinctions of known transitions, such as that at 590 nm (Fe^{II}–CO; cf. Fig. 6.3 and Clore et al., 1980b). Chance and Leigh (1977) quote a much higher extinction, but it is not clear how it was determined. Compound C also has a weak β-band in the 550–570 nm region (Clore et al., 1980b; Denis, 1981; Nicholls, 1979b). It exhibits no specific EPR resonances from the a_3/Cu$_B$ centre (Chance et al., 1975; Clore et al., 1980b), but the EPR spectrum shows clearly that both haem a and Cu$_A$ remain fully oxidized.

Haem iron of a_3 has generally been proposed to be ferrous in Compound C (but see Erecińska and Wilson, 1978). Chance et al. (1979) and Chance and Leigh (1977) suggested the configuration

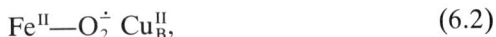

$$Fe^{II}\!-\!O_2^{\pm}\ Cu_B^{II}, \qquad (6.2)$$

ascribing the strong 605 nm band to charge transfer between iron, copper and the superoxide radical. Powers et al. (1979) proposed alternatively that this band may be due to Cu$_B^{II}$ being a "Type 1" copper (cf. Malkin and Malmström, 1970, for nomenclature), the absorption band of which may become unmasked due to interaction between haem iron, copper and $O_2^{\pm}$. However, as pointed out by Clore et al. (1980b), this band is much narrower than expected from a copper transition. Instead, the latter authors proposed the configuration

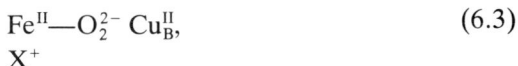

$$Fe^{II}\!-\!O_2^{2-}\ Cu_B^{II}, \qquad (6.3)$$
$$X^+$$

with oxygen reduced to peroxide and X$^+$ standing for either a porphyrin or a protein free radical.

The absence of the 655 nm band typical of the fully oxidized enzyme from Compound C has been used as an argument for a ferrous state of haem a_3 (Denis, 1977, 1981; Chance and Leigh, 1977; Clore et al., 1980b). However, this does not exclude a ferric state because the 655 nm band has been shown to oe absent in some states of the enzyme where haem a_3 can be proved to be ferric by EPR (Shaw et al., 1978). The strong 605 nm band is consistent with a low spin ferrous state. However, from difference spectra it appears that the Soret band is much weaker and the α-band much stronger than those of the low spin ferrous CO compound (Chance et al., 1979; Denis, 1981). Thus, in comparison with the CO compound, Compound C has an enhanced α-band but a decreased Soret band. The latter cannot be much more extensive than the band of the fully oxidized enzyme, but is shifted towards the red.

The structures in (6.2) and (6.3) may also be difficult to accept following the hypothesis (Section II) that the binuclear centre has evolved to facilitate concerted two-electron transfer.

We must also consider the possibility that CO may remain bound to Cu_B also after the Compound A stage (Section IV.A). In fact, Nicholls (1978, 1979a,b) specifically proposed that this is the case in Compound C so that this species would be an artefact of the low temperature technique in which the use of CO is essential. However, this possibility was recently excluded by the finding (Wikström, 1981c) that Compound C may also be generated from the fully oxidized enzyme by ATP-linked reversed electron transfer in the complete absence of CO (cf. Chapter 7). The fact that Compound C is generated under these highly oxidizing conditions directly contradicts the possibility that haem a_3 is in the ferrous state.

The proposals of a Cu_B^{II} state in Compound C do not have particularly strong experimental support. This is mainly because no optical transition has so far been unambiguously shown to represent Cu_B^{II}. The weak 740 nm band (Chance and Leigh, 1977) and other changes in the near infrared region (see Clore et al., 1980b) might also be attributable to haem transitions (see also Eglinton et al., 1980).

Two further alternatives may be considered for the structure of Compound C. The first is a μ-peroxo species of ferric haem iron and Cu_B^{II} (cf. Erecińska and Wilson, 1978; Caughey et al., 1979; Wikström, 1981c; Wikström et al., 1981), viz.

$$Fe^{III}-O^--O^--Cu_B^{II} \quad \text{or} \quad Fe^{III}-\underset{\underset{O^-}{|}}{O^-}-Cu_B^{II} \qquad (6.4)$$

The extensive 605 nm band might then be due to charge transfer from the axial ligand of high electron density (Brill and Williams, 1961; George et al.. 1961; Yamazaki et al., 1966). Antiferromagnetic coupling between Fe^{III} and Cu^{II} may account for the absence of specific EPR resonances. Such a species would indeed fit several of the requirements discussed above. However, the optical spectrum is, no doubt, quite unusual for a ferric haem.

For this reason we have more recently considered another alternative (Wikström, 1981d), viz. the species in (6.5) with ferryl haem iron and cuprous copper:

$$Fe^{IV}=O^--O^- \; Cu_B^I \qquad (6.5)$$

This proposal is related to the suggestion by Seiter and Angelos (1980) that haem a_3 may cycle between Fe^{II}, Fe^{III} and Fe^{IV} states during catalysis. We do not, however, concur with their proposal that Cu_B does not undergo oxidoreduction for reasons discussed in Chapter 4. The Fe^{IV} state may well yet be consistent with oxidoreduction of Cu_B, as suggested below (Section V).

Chance et al. (1975) also considered briefly the possibility of a Fe^{IV} state in Compound C. Clore et al. (1980b) discounted this possibility on the basis that Compound C is spectrally dissimilar to known Fe^{IV} haemoproteins. However, this apparent dissimilarity may simply be due to the fact that the haem is iron protoporphyrin IX in cytochrome c peroxidase (Yonetani, 1970), horseradish peroxidase (George, 1953) and myoglobin (George and Irvine, 1955) while it is haem A in cytochrome oxidase. The electron-withdrawing properties of the formyl group in haem A would be expected to shift the spectrum to the red to a considerable extent. When this is taken into account the spectrum of Compound C is indeed rather similar to the ES complex of cytochrome c peroxidase (Yonetani, 1970), horseradish peroxidase Compound II (George, 1953) and myoglobin (Fe^{IV}; see, for example, Churg et al., 1979). Thus Compound C exhibits a β-band and a Soret band that is red-shifted but of similar magnitude to the corresponding ferric species. The strong α-band (Q_0-band; see Churg et al., 1979) as compared to the much weaker β- (or Q_v-) band in Compound C could be related to asymmetry imposed by the peroxo ligand, as opposed to the more symmetrical $Fe^{IV}{=}O$ bond of ferryl metmyoglobin (Churg et al., 1979).

One further analogy might be appropriate here. The spectra of the Fe^{IV} compounds of peroxidases and myoglobin are remarkably similar to the spectrum of ferrocytochrome b (also iron protoporphyrin IX). If this can also be taken more generally to suggest that the spectrum of ferryl haem iron compounds may be similar to the corresponding low spin ferrous state, it would support the Fe^{IV} state in Compound C. The optical spectrum of the latter is indeed similar to that of ferrocytochrome a (or the CO compound of a_3) with the exception of the above-mentioned anomalies of the Soret band.

The species in (6.5) is not expected to show any EPR resonances, in agreement with experiment.

Clore et al. (1980b) and Denis (1981) showed that Compound C may really be two species that are formed sequentially after Compound A. However, their spectral difference is small. We agree with Denis (1981) that this may be attributed to a slight conformational change. This may, for example, be due to dissociation of CO off the copper (cf. above), or to formation of a hydrogen bond between peroxide and a protonated group in the haem pocket (cf. Caughey et al., 1979).

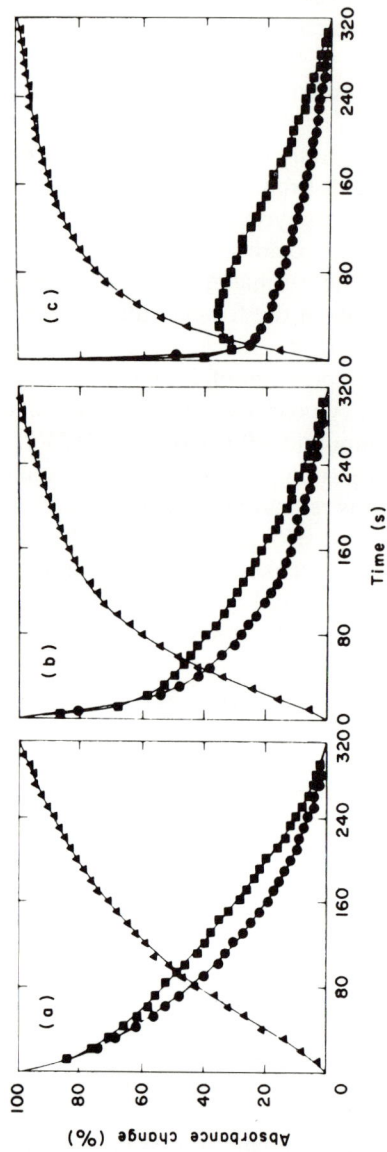

Fig. 6.4 Kinetics of oxidation of reduced cytochrome oxidase with O_2 at three wavelength pairs. Reaction temperature 176 K. ●——●, 604–630 nm; ■——■, 608–630 nm; ▲——▲, 830–940 nm. The O_2 concentrations were 60, 200 and 1180 μM in (a), (b) and (c), respectively. From Clore and Chance (1978a) with permission.

C. The nature of Compound B

If Compound C is indeed a "peroxy" intermediate of catalytic importance, it should also form in the reaction between the fully reduced enzyme and O_2. However, previously this reaction has been described to be profoundly different from the reaction of the half-reduced enzyme (Chance *et al.*, 1975; Clore *et al.*, 1980*a*; see also Fig. 6.3).

When the fully reduced enzyme reacts with O_2 a state termed Compound B is formed (Fig. 6.3). However, Clore *et al.* (1980*a*) and Clore and Chance (1978*a*) showed both kinetically (Fig. 6.4) and by consecutive

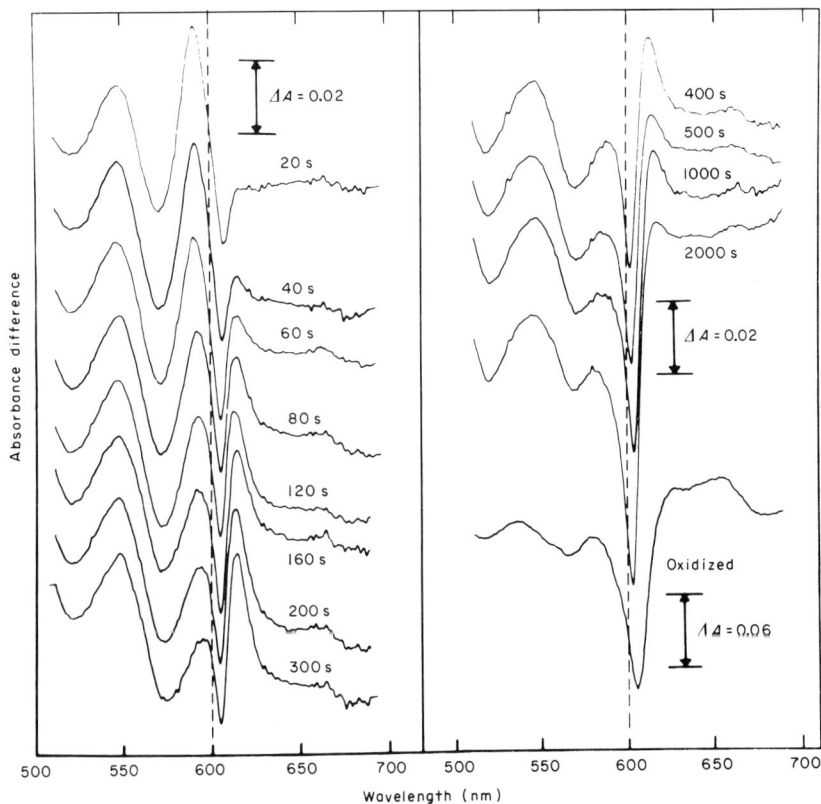

Fig. 6.5 Spectral analysis of the reaction of the reduced enzyme with O_2. Reaction temperature 173 K. Consecutive difference spectra (relative to the fully reduced enzyme after dissociation of CO) are shown. A difference spectrum of fully oxidized minus fully reduced enzyme (denoted "Oxidized") is also shown for comparison. Traces are labelled with the reaction time from flash-photolysis in seconds. From Clore *et al.* (1980*a*) with permission.

spectra (Fig. 6.5) a clear intermediate between Compounds A and B, which they called "II" (Compounds A and B were called I and III, respectively). The main characteristic of "II" is that its formation from Compound A is associated with an *increase* in absorption in the 605–610 nm region (see Fig. 6.4(c) and Fig. 6.5, traces 60–300 s). We find it likely to be more than a coincidence that the conversion of Compound A to the next intermediate should be associated with an *a priori* unexpected absorption increase in the 605 nm region both when the fully reduced and the half-reduced (Section IV.B) enzyme reacts with dioxygen. We therefore suggest that the intermediate "II" is indeed equal to the "peroxy" Compound C (Section IV.B), except that haem a and Cu_A are reduced in the former but oxidized in the latter case. The transient nature of "II", as opposed to Compound C, would consequently be expected to be the result of electron transfer from haem a and Cu_A which leads to the state termed Compound B.

Haem a and Cu_A are indeed partially oxidized in Compound B (see Erecińska and Wilson, 1978; Clore *et al.*, 1980*a*), although this has not been more generally appreciated. However, this is proven unambiguously by the 30–50% appearance of the $g = 3$ EPR signal (Chance *et al.*, 1975; Clore *et al.*, 1980*a*) relative to the fully oxidized enzyme, and by the optical difference spectrum (Fig. 6.3). Also Cu_A is oxidized to about the same extent, as revealed by EPR (Chance *et al.*, 1975) and by the increase of the 830 nm band (Chance and Leigh, 1977). Complete oxidation of Cu_A, as proposed by Clore *et al.* (1980*a*), contradicts the earlier data, and this point may require further clarification.

These results lead to the conclusion that haem a and Cu_A become *partially* oxidized in Compound B, which means that this state of the enzyme is heterogeneous (cf. Clore *et al.*, 1980*a*). The remainder of cytochrome a is oxidized only at much higher temperatures, and then in conjunction with oxidation of cytochrome c (Chance *et al.*, 1978). We consider this kinetic heterogeneity to be an important clue to the electron transfer mechanism in cytochrome oxidase, and this point will be considerably developed below (Chapter 7). First, we should ascertain that this *a priori* unexpected behaviour is not an artefact of the low temperature technique.

We are now in the position to compare the low temperature kinetics with the kinetics of the reaction at room temperature (Section III). It may first be noted that the heterogeneous or biphasic oxidation of cytochrome a deduced from the low temperature experiments provides a long-sought-for explanation of the room temperature kinetics. The reason for the large extent of the fast phase of oxidation at 605 nm (cf. Fig. 6.1(a), Section III) is clearly that it includes both oxidation of haem a_3 and the 30–50% oxidation of haem a (cf. above). Similarly, the biphasic oxidation of Cu_A is also observed at room temperature (Fig. 6.1(a)). Interpretation of the absorption changes at 830 nm is no longer subject to ambiguity (see Beinert *et al.*, 1980; Boelens and Wever, 1980).

Moreover, the transient formation of intermediate "II" of Clore *et al.* (1980*a*; cf. above), was presumably also observed earlier at room temperature (Fig. 6.1(b)), and is ascribed here to the transient appearance of the "peroxy" intermediate.

In conclusion, the kinetics of the cytochrome oxidase reaction, including observed intermediates, are quite similar at high and low temperatures. This is rewarding because it confirms the usefulness of the "triple trapping" technique, which presently is the only method that promises to give a mechanistic insight into the dioxygen reaction. We also conclude that the reaction of fully reduced and half-reduced enzyme with O_2 may be much more similar than anticipated previously, which also tends to simplify interpretation. Finally, the oxidations of cytochrome a and of Cu_A exhibit a kinetic heterogeneity that on first sight appears anomalous. However, in Section VI it will be shown that a similar heterogeneity may also be deduced from reductant-pulse experiments. A functional rationalization of this "anomaly" will be proposed in Chapter 7.

D. The intermediates beyond Compound B

Low temperature O_2 pulse studies of cytochrome oxidase have basically been carried out under two quite different sets of conditions, which must be constantly remembered when comparisons are made between data from different laboratories. Thus most of the low temperature experiments cited above have been carried out in the presence of ferrocytochrome c, either in intact mitochondria or with the isolated enzyme. In contrast, the more recent low temperature studies which have revealed intermediates with new EPR signals either from haem (Shaw *et al.*, 1979; cf. Chapter 4, Section III.B) or from Cu_B (Karlsson and Andréasson, 1981) have been performed in the absence of an extraneous electron donor. This latter condition is comparable to the studies on the different "oxygenated" forms of the enzyme, carried out for example by Orii and King (1972, 1976). The absence of ferrocytochrome c is expected to slow down considerably late events in the enzyme's catalytic cycle, possibly making "late" intermediates stable for long periods of time.

Shaw *et al.* (1979) described an oxygen intermediate with a unique EPR signal at $g = 5$ (Chapter 4, Section III.B) following oxygenation of the fully reduced enzyme. This species appeared later than the Compound B state, and seemed to coincide kinetically with either "Compound I" or "Compound II" described earlier by Orii and King (1972, 1976). (Note that these "Compounds" are not the same as Compounds I–III described by Clore *et al.* (1980*a,b*) and discussed above.) "Compound I" is spectrally similar to Compound C (cf. Erecińska and Wilson, 1978), but in the spectra of Orii and King (1972) it already appeared to have a considerable contribution from the kinetically subsequent "Compound II". The latter is characterized by a red-shifted Soret band relative to the fully oxidized "resting" enzyme, and by a broad absorption peak near 580 nm. According to the reflectance spectra of Shaw *et al.* (1979) and the absence of a $g = 5$ signal from Compound C, it seems likely that the $g = 5$ species is identical with "Compound II" of Orii and King. It is also noteworthy that the $g = 5$ species exhibits the 655 nm band, a property that it therefore shares with the "resting" enzyme.

The work by Karlsson and Andréasson (1981), in which they studied the newly discovered EPR-detectable form of Cu_B^{II} (cf. Chapter 4), was performed under rather similar conditions. It is interesting that after oxygenation of the reduced enzyme they observed that broad absorption near 580 nm (cf. above) coincided with the new EPR signal of Cu_B^{II}, and that when both decayed in parallel the 655 nm band was formed with similar kinetics. There is hence a slight discrepancy to the results of Shaw *et al.* (1979), who in their $g = 5$ state observed the 580 and 655 nm absorptions simultaneously. However, this may simply be due to some overlap between subsequent intermediate states. Our impression is that the EPR-detectable state of Cu_B^{II} may be formed first, followed by generation of the state characterized by the $g = 5$ signal (and the 655 nm absorption).

Unequivocal assignments of the valence states of the haem a_3/Cu_B centre are difficult for these "late intermediates". However, recent data on energy-dependent partial reversal of the oxidase reaction (Wikström, 1981*c*; see also Chapter 7) may provide some clues. In this work it was shown that the fully oxidized enzyme may be converted to a species with a broad absorption near 580 nm (relative to the "resting" state) in a reversed one-electron transfer process coupled to the hydrolysis of ATP. The electron donor is probably water or its equivalent, and the acceptor ferricytochrome *c*. The product would thus be an one-electron oxidation product of the fully oxidized centre plus water. Moreover, the reversed electron transfer was also shown to proceed one electron step further at a higher driving force (ATP/ADP·P_i ratio or redox potential at cytochrome *c*), by which the haem a_3/Cu_B centre is converted into a state spectrophotometrically similar to Compound C (Wikström, 1981*c*). Formation of the 580 nm state was already associated with decrease of the 655 nm absorbance.

Hypothesizing from this information we may consider the possible results of transferring one electron from cytochrome *a* to the a_3/Cu_B centre, where the latter is in the "peroxy" or Compound C configuration. We suggest tentatively that whether the peroxy state has the structure in (6.4) or that in (6.5), the result may be a concerted two-electron reaction (see also Section II), the second electron originating either from a_3 (6.4) or from Cu_B^I (6.5) with generation of the structure in (6.6) or its electronic equivalent.

$$Fe^{IV}=O^{2-} \ Cu_B^{II}-OH^- \tag{6.6}$$

This configuration may well result in EPR-visibility of the copper, as observed by Karlsson and Andréasson (1981; cf. Chapter 4), but with haem iron being non-detectable by EPR. In the next stage this species may undergo a rearrangement such as to introduce antiferromagnetic coupling between haem iron and copper. If this occurred before arrival of the last electron, it could result in a binuclear $S = \frac{3}{2}$ centre with the unique EPR properties described by Shaw et al. (1979); cf. Chapter 4) and also in formation of the 655 nm band characteristic of the coupling between copper and haem iron. Simultaneously both the 580 nm band and the EPR signal of Cu_B^{II} would be lost. We may add here that although the 655 nm band has been attributed to ferric haem a_3 in its special linkage with copper, it is not excluded that this band may be due to Cu_B^{II} in special linkage with haem a_3.

Finally, it must be stressed that all structures proposed here are tentative, requiring verification or disproval in further studies. It seems particularly desirable to obtain MCD and magnetic susceptibility data on trapped "oxygen intermediates" at low temperatures.

V. Tentative catalytic mechanism of O_2 reduction

From the preceding sections we now have the main premises to propose how O_2 reduction might take place at the haem a_3/Cu_B centre. A tentative proposal is shown in Fig. 6.6. Even though this scheme is hypothetical, it is simple and consistent with the main experimental findings.

Figure 6.6 is really almost self-explanatory and requires little discussion beyond that of the previous sections. It is significant that reduction of O_2 takes place in two concerted two-electron steps, and yet that electron transfer to the a_3/Cu_B centre occurs by discrete one-electron steps. The latter feature is considered important for the coupling of oxidoreduction of cytochrome a to proton translocation (Chapter 7).

The fully oxidized centre might exist in at least two isoelectronic configurations. Of these the ferryl iron (no. 3, Fig. 6.6) and μ-oxo (Chapter 4) states might correspond to "pulsed" (or "oxygenated") and "resting" states of the enzyme, respectively (Chapter 4 and Section VI.B).

Intermediate no. 4 (Fig. 6.6) may contain high spin ferric haem and

$$Fe^{II} \mid Cu^{I}$$
$$\downarrow O_2$$
$$Fe^{II}\text{-}O_2\ Cu^{I}(CO\,?)$$
$$\downarrow$$

1 $\boxed{Fe^{III}\text{-}O^{-}\text{-}Cu^{II} \text{ or } Fe^{IV}=O^{-}\text{-}O^{-}Cu^{I}, \ O^{-}}$

$e^- + H^+ \longrightarrow \qquad H_2O \longleftarrow \quad \longleftarrow e^- + H^+$

2 $\boxed{Fe^{IV}=O^{=}Cu^{II}\text{-}OH^{-}}$ $\qquad\qquad$ $\boxed{Fe^{III}\text{-}OH^{-}Cu^{I}(O_2)}$ **4**

$e^- + H^+ \longrightarrow \quad \longrightarrow H_2O \qquad\qquad \longleftarrow e^- + H^+$

3 $\boxed{Fe^{IV}=O^{=}Cu^{I}}$ $\qquad O_2$

$$\downarrow$$
$$Fe^{III}\text{-}O^{=}\text{-}Cu^{II}$$

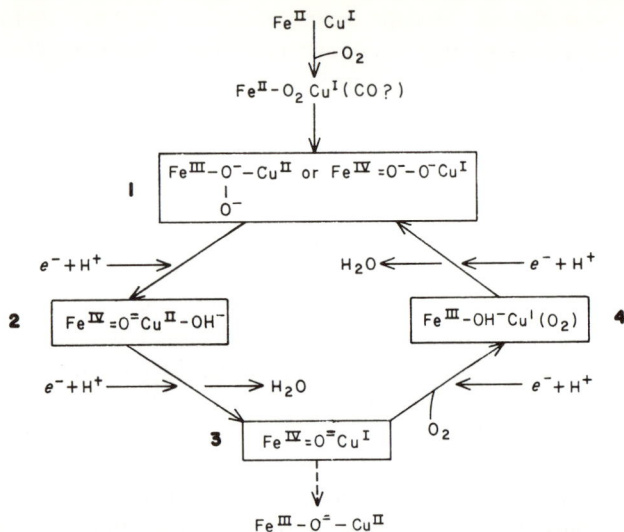

Fig. 6.6 Tentative mechanism of O_2 reduction at the haem a_3/Cu_B centre. Intermediate **1**: peroxidic species, either in the μ-peroxo configuration, or with quadrivalent haem iron. Intermediate **2**; water oxygen bound to ferryl haem iron. Intermediate **3** (fully oxidized state); ferryl haem iron or μ-oxo configurations. Intermediate **4**: high spin ferric haem with O_2 bound to reduced copper. Electron input from cytochrome a is shown, as well as the uptake of H^+ required in the generation of water. The centre may never be fully reduced during catalysis, as shown. Alternatively, a Fe^{II}—O_2 Cu^{I} species might be a short-lived intermediate between **4** and **1**.

should be detectable by EPR spectroscopy. The rhombic signal at $g = 6$ observed by Rosén et al. (1977) under turnover conditions might stem from this type of intermediate (see also Karlsson and Andréasson, 1981). Also in some other respects our scheme resembles that proposed by these workers. In making comparisons one should note that the scheme in Fig. 6.6 is much simplified in that several possible intermediate states are omitted and the reactions shown are therefore often composite. One main difference is that we have considered states with bound O_2^- and O^- radicals unlikely as intermediates with significant occupancy during turnover, thereby introducing the concept of concerted two-electron transfer and involvement of the ferryl state. Another stems from the recent data on reversal of the oxygen reaction (Wikström, 1981c) which indicate that the 580 nm state may be an intermediate before the fully oxidized state rather than after in the forward reaction. However, EPR studies of the states generated by reversed electron transfer are badly needed to test our proposal.

The experiments on the reversal of the oxidase reaction (Wikström,

1981c) suggest that it may be the formation of the "peroxy" intermediate (no. **4** → no. **1**; Fig. 6.6) that provides the irreversibility of the overall catalytic cycle. Whether this is due to a particularly negative free energy change of this step, or whether it is a mechanistic consequence is not known.

As discussed in Section IV.A, the "oxy" Compound A might accumulate artificially in low temperature experiments where CO is employed. The possibility must be considered (as drawn in Fig. 6.6) that copper is the primary acceptor of O_2, as proposed by Erecińska and Wilson (1978). This is consistent with the findings of Alben et al. (1981; Section IV.A) and also with the NO-binding data (Chapter 4). However, we cannot of course exclude the possibility that the Fe^{II}—O_2 compound might be a short-lived intermediate between the species nos **4** and **1** in Fig. 6.6.

Binding of CO to Cu_B^I might provide an explanation for the finding by Nicholls (1978, 1979a,b) that a species very similar to Compound C is generated from the oxidized enzyme simply by treating the enzyme suspension with CO. This is our main reason for proposing that the oxidized centre might take the electronic configuration of no. **3** in Fig. 6.6. In the fully oxidized enzyme this state could exist in thermal equilibrium with other, more dominant, species such as the μ-oxo form. However, it may be stabilized by binding of CO to the copper, after which mobilization of endogenous reductants often present in oxidase preparations may result in generation of Compound C or its analogue in a reaction with O_2.

A perhaps more interesting possibility is brought forward by the finding that cytochrome oxidase catalyses slow oxidation of CO to CO_2 in a reaction that is accelerated by O_2 (Young et al., 1979). Thus

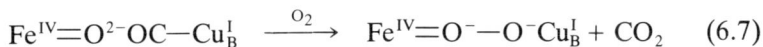

$$Fe^{IV}{=}O^{2-}OC{-}Cu_B^I \xrightarrow{\ O_2\ } Fe^{IV}{=}O^-{-}O^-Cu_B^I + CO_2 \quad (6.7)$$

Also in this case the reaction would be initiated by stabilization of the ferryl iron/cuprous state (no. **3**; Fig. 6.6) by CO. The product of (6.7) can, of course, also be written in the form of (6.4).

VI. Reduction of cytochrome oxidase by cytochrome c

A. Electron transfer between cytochromes c and a

When isolated and purified cytochrome oxidase is pulsed with ferrocytochrome c under aerobic or anaerobic conditions, there is fast electron transfer ($k = 8 \times 10^6$ M^{-1} s^{-1}) from the latter to the former (Gibson et al., 1965; Andréasson et al., 1972; Van Buuren et al., 1974; Andréasson, 1975; Wilson et al., 1975). The initial burst of electron transfer is unaffected by either cyanide or azide, and also occurs in the "mixed-valence" state (Greenwood *et al.*, 1976). Thus, electrons do not reach the haem a_3/Cu_B centre in the fast burst, but only much later. This slowness of electron transfer between cytochromes a and a_3 is a long known characteristic of the fully oxidized "resting" enzyme. Electron transfer is clearly blocked transiently, and normal turnover ensues only relatively slowly (see Section VI.B).

The interpretation of the redox centre events during the fast burst has been the subject of some controversy. Andréasson *et al.* (1972) and Andréasson (1975) found that no more than one electron equivalent per aa_3 unit is transferred from cytochrome c to the oxidase during the fast burst, no matter how high the applied c/aa_3 ratio. This was interpreted to mean that only one acceptor is reduced in the oxidase, viz. haem a. However, the other groups cited above, while agreeing that haem a is the primary electron acceptor, proposed that Cu_A is also reduced in the fast burst, both centres becoming roughly half-reduced. This is supported by fast disappearance of part of the 830 nm band as well as of the $g = 2$ signal (Van Gelder and Beinert, 1969; Beinert *et al.*, 1976). This must now be considered strong evidence for reduction of Cu_A (cf. Chapter 4). Moreover, the calculated extinction coefficient for reduction at 605 nm is about 13 mM^{-1} cm^{-1} if haem a is assumed to receive all electrons donated by cytochrome c (Andréasson, 1975). Since it can now be shown that about 80% of the 605 nm band is due to haem a (Chapter 4), this also shows that only some 60% of the donated electrons reduce haem a, in good agreement with the conclusions of Wilson *et al.* (1975).

Although this means that we must take a stand against the interpretation by Andréasson *et al.* in the above sense, it does not retract from their important finding. In fact, the data of Wilson *et al.* (1975) also show that the oxidase accepts no more than one e^- per aa_3 unit no matter how high the employed c/aa_3 ratio, although this point was not made explicit in the latter work. It is thus clear that the difference between these workers is not of experimental origin.

The conclusion from this work must be that cytochrome oxidase accepts no more than one e^- per aa_3 unit in the fast burst of electron transfer, but that the accepted electrons are distributed about equally between haem a and Cu_A. The important question is clearly why the enzyme cannot accept two e^- per aa_3 unit even though both haem a and Cu_A appear to be "active" electron acceptors. The first possibility is that a tight complex is formed between ferric cytochrome c and the oxidase (see Section VI.B), preventing transfer of a second electron. However, this was excluded by Andréasson (1975), who demonstrated that the second-order rate constant for the reaction between ferrous c and the oxidase was unaffected by ferric cytochrome c. The second possibility is a strong negative redox interaction between haem a and Cu_A, making simultaneous reduction of both unlikely. This explanation is not supported by available redox data (Chapter 5), in which there is no indication of such an interaction. A third possibility is highly interesting. Reduction of haem a and Cu_A may be possible in only half of the aa_3 monomers present, i.e. a so-called "half-of-the-sites" effect. This would suggest functioning of the enzyme as a dimer with strong antico-operativity between monomers, an idea that is further developed in Chapter 7.

When the fully reduced enzyme is pulsed with ferricytochrome c, electron transfer is observed from the former to the latter (Petersen and Andréasson, 1976). In contrast to the opposite experiment described above, this results in transfer of maximally *two* electrons per aa_3 unit, by which the oxidase goes to a state spectroscopically similar to the half-reduced situation during anaerobic potentiometric titrations (Chapter 5). The apparent E_m for this transition is also comparable to the equilibrium situation.

B. Steady state kinetics and mechanism of the cytohrome c reaction

Although cytochrome c is one of the best characterized proteins (see Dickerson and Timkovich, 1975), the mechanism by which it interacts with the oxidase and by which electrons are transferred is not yet fully understood. It now seems to be established that there are two binding sites for cytochrome c on the oxidase of high and low affinity, respectively, although further binding may occur to phospholipid present in the preparation. The high affinity site is on subunit II, to which cytochrome c binds with a region involving seven lysine residues. The low affinity site seems to represent binding to cardiolipin tightly associated to the oxidase (Chapter 3).

Early work by Smith and Conrad (1956) revealed an apparent paradox, viz. that when the oxidase reaction was followed polarographically in the presence of cytochrome c plus ascorbate as reductant, it displayed Michaelis–Menten kinetics with respect to cytochrome c, but was first order with time even at high c concentrations when measured spectrophotometrically in the absence of ascorbate. Minnaert (1961) explained this by his "Mechanism IV", which has subsequently been generally adopted as a basis for further more detailed mechanisms. "Mechanism IV" may be characterized by the following reactions:

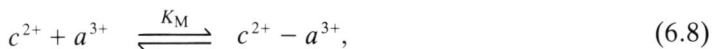

$$c^{2+} + a^{3+} \quad \underset{\longleftarrow}{\overset{K_M}{\rightleftharpoons}} \quad c^{2+} - a^{3+}, \tag{6.8}$$

$$c^{2+} - a^{3+} \quad \longrightarrow \quad c^{3+} - a^{2+} \quad \xrightarrow{O_2} \quad c^{3+} - a^{3+}, \tag{6.9}$$

$$c^{3+} - a^{2+} \quad \overset{K_I}{\rightleftharpoons} \quad c^{3+} + a^{3+}. \tag{6.10}$$

Subsequently, formation of "tight" complexes between the oxidase and cytochrome c was demonstrated with binding constants comparable to the kinetic constants (see Nicholls and Chance, 1974), K_M and K_I characterizing the equilibria of reactions (6.8) and (6.10), respectively. Although it seems reasonable to assume that the formation of such complexes is of significance in the reaction mechanism, this question is not yet solved unambiguously. The alternative explanation is that these complexes are inactive catalytically ("dead end") and that electron transfer takes place via transition state complexes of very short lifetime (see Errede and Kamen, 1979).

As pointed out by Nicholls (1974c), Minnaert's "Mechanism IV" is compatible with the observed first-order kinetics only when the affinity of the oxidase is the same for reduced and oxidized cytochrome c ($K_M = K_I$; see equations (6.8) and (6.10)). This contrasts to binding studies (see Vanderkooi and Erecińska, 1976), which suggest tighter binding of ferric than of ferrous cytochrome c. However, this may be due to a preferred binding of the ferric form to phospholipid whereas binding to the specific sites could remain equally strong for both species. But, as has been pointed out (Nicholls, 1974c; Vanderkooi and Erecińska, 1976), endogenous cytochrome c of mitochondria and submitochondrial particles has been reported to have an E_m about 60 mV lower than the E_m in solution (Dutton $et\ al.$, 1970), suggesting considerable preference for binding of the ferric form to the specific sites of the oxidase (see also Schroedl and Hartzell, 1977c). This means that there is a discrepancy between "Mechanism IV" and the binding data that may be difficult to explain. Very recently Denis et $al.$ (1980) reported data that may resolve this problem. By using a spectrophotometric/potentiometric technique with careful control of light-scattering changes these authors reported that the E_m of cytochrome c in yeast mitochondria is almost identical to E_m of the free species (i.e. 285–290 mV under their experimental conditions). These authors attribute the apparent lowering of E_m upon binding of cytochrome c to an artefact that is due to light scattering. If this claim can be confirmed, it would, of course, immediately solve the problem. It would also, incidentally, have an appreciable effect on the oxidoreduction potential estimated for cytochrome c in aerobic steady states (see Wikström and Krab, 1979), and on apparent E_m values of cytochrome a estimated under similar conditions (Wikström $et\ al.$, 1976).

More recent extensive kinetic data (Errede $et\ al.$, 1976; Ferguson-Miller $et\ al.$, 1976) have suggested that "Mechanism IV" may have to be modified to take into account binding of two cytochrome c molecules per aa_3 unit during catalysis. Appropriate rate laws for mechanisms in which the two sites are dependent or independent have been derived by Errede and Kamen (1978; see also Errede $et\ al.$, 1979), and a distinction between the two is presently not possible on the basis of steady state data. Negative co-operativity of binding to the two sites was proposed by Ferguson-Miller *et al.* (1976, 1978*b*) and may explain the finding of Bisson *et al.* (1980) that covalent binding of a *c* derivative to the high affinity site, leaving the low affinity site free for binding, still blocks electron transfer activity (see Chapters 3 and 7).

The effect of chemical modification of lysyl groups on cytochrome *c* on activity and on binding to both the oxidase and to cytochrome *c* reductase (i.e. cytochrome bc_1 complex), as well as to isolated cytochrome c_1 (Ferguson-Miller *et al.*, 1979; Rieder and Bosshard, 1980), has indicated strongly that cytochrome *c* binds to the c_1 portion of the bc_1 complex with the same binding region that is utilized in the high affinity binding to the oxidase. This was interpreted to mean that a "solid state" link between reductase and oxidase by cytochrome *c* is unlikely, but that it must undergo at least some rotational motion to carry electrons from reductase to oxidase. In contrast, Smith *et al.* (1973) suggested different areas on the surface of cytochrome *c* to be involved in the reaction of cytochrome *c* with oxidase and reductase. These workers found that an antibody against cytochrome *c*, forming a 1 : 1 complex with the latter, blocked the reaction with the oxidase but not with the reductase. In the light of the more recent data these most interesting findings require re-interpretation. Thus they may suggest, more generally, that the reaction of cytochrome *c* with oxidase and reductase is also dependent on structural features of either the oxidase or the *c* itself other than the area of seven lysyl residues establishing the contact. Either the binding of cytochrome *c* to the oxidase is dependent on such extra factors (which are different from those of the reductase), or then the difference should be ascribed to the catalytic mechanism involving the bound cytochrome *c*. Unfortunately Smith *et al.* (1973) did not determine whether antibody-complexed *c* bound to the oxidase. If binding does take place, several possibilities remain, many of which are amenable to experimental test. Such experiments may be encouraged because they may well reveal important information on the mechanism of the cytochrome *c*–cytochrome oxidase interaction. It could be speculated, for instance, that the oxidase functions as the dimer (see Chapter 7) and requires binding of two cytochromes *c* per molecule (i.e. one *c* per aa_3), which may not be possible with the large antibody–cytochrome *c* complex for steric reasons.

C. "Resting" and "O_2-pulsed" states of the enzyme

If the fully reduced enzyme is pulsed with O_2 in the presence of ferrocytochrome *c*, electron transfer from the latter (following the very fast initial phase, see above) is stimulated four- to fivefold relative to the situation in the "resting" enzyme (Antonini *et al.*, 1977; Rosén *et al.*, 1977; Wilson *et al.*, 1978; Brunori *et al.*, 1979). From this it was inferred that the enzyme does not reach the catalytically inferior "resting" state after the reaction with O_2 but a catalytically more potent "O_2-pulsed" state, in which there is much less or no restriction towards electron transfer between cytochromes *a* and a_3. Comparable catalytically more active states are also reached by preincubation of the enzyme in the presence of substoicheiometric amounts of ferrocytochrome *c* (causing partial reduction; Rosén *et al.*, 1977), and also subsequent to fast anaerobic oxidation of the fully reduced enzyme by porphyrexide (Shaw *et al.*, 1978; Chapter 4). Thus the kinetic limitation typical of the "resting" state may be removed by partial reduction of the enzyme prior to the pulse of reductant. In principle, this might also be the basic reason for the high catalytic competence of the "pulsed" state. Even though it contains four oxidizing equivalents overall per aa_3 (Chapter 4), a partially reduced form of O_2 is likely to remain bound to the a_3/Cu_B centre so that reducing equivalents may still be located in the redox centres. However, as also shown by Brunori *et al.* (1979), the "pulsed" state is not the prevailing one during the aerobic steady state, in which the spectrum resembles more that of the fully oxidized form.

Another possible structure of the "pulsed" state is the ferryl haem iron species no. **3** in Fig. 6.6, which represents an isoelectronic variant of the "resting" μ-oxo state. In the absence of electron donors, the enzyme may relax to the latter state, but in their presence other intermediate states besides the fully oxidized one (no. **3**, Fig. 6.6) may also be responsible for the high catalytic competence. The rhombic $g = 6$ signals observed in such partially reduced and kinetically competent states (Rosén *et al.*, 1977; Shaw *et al.*, 1978) could thus arise, for instance, from states analogous to no. **4** (Fig. 6.6).

Finally, the kinetic heterogeneity in oxidation of cytochrome *a* should also be taken into account in considerations of possible structures of the "pulsed" enzyme. The data of Brunori *et al.* (1979) do not rule out the possibility that the "pulsed" state is inhomogeneous with different states of aa_3 units blended together. In our view it was also not ruled out that part of haem *a* may be reduced in this state (cf. Chapter 4). Altogether it is clear that freeze-trapping of the "pulsed" state with thorough spectroscopic analysis may be necessary to test the many possibilities that still remain. In Chapter 7 (Section VI.C) we will discuss a further possible explanation of the "pulsed" state.

D. Electron transfer sequence

The haem of cytochrome *a* is no doubt the primary electron acceptor of the enzyme (Section VI.A). From the reductant-pulse experiments it seems equally clear that Cu_A accepts electrons next. It is findings such as these that have led to the assumption that the (linear) electron transfer sequence in the oxidase is

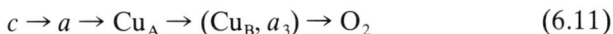

$$c \rightarrow a \rightarrow Cu_A \rightarrow (Cu_B, a_3) \rightarrow O_2 \qquad (6.11)$$

However, the discovery of heterogeneous (or biphasic) kinetics of haem a and Cu_A oxidation and reduction has made us strongly doubt this simple sequence. Room temperature kinetics, for example (Fig. 6.1(a)), show that the fast phase of haem a oxidation occurs prior to the fast phase of Cu_A oxidation, and that both these events are much faster than the rates of oxidation of the remaining parts of both haem a and Cu_A (which are kinetically indistinguishable).

The following schematic sequence seems to be more consistent with these data:

$$c \longrightarrow \begin{array}{c} (a) \\ \updownarrow \\ Cu_A \end{array} \longrightarrow \begin{array}{c} (a) \\ \updownarrow \\ Cu_A \end{array} \longrightarrow (Cu_B, a_3) \longrightarrow O_2 \quad (6.12)$$

The possible significance of the two forms of cytochrome a that we find it necessary to postulate is discussed further in Chapter 7. We should stress, however, that there is to our knowledge no evidence for the idea that Cu_A is an obligatory intermediate in the electron transfer sequence since it is reduced later than haem a on fast reduction and oxidized later than haem a on fast oxygenation.

VII. Conclusion

A tentative mechanism of O_2 reduction was proposed, which is in agreement with theoretical predictions as well as with spectroscopic and kinetic findings.

Kinetic anomalies are found in the oxidation and reduction of both cytochrome a and Cu_A. These cannot be attributed to experimental artefacts, but are likely to be important characteristics of the catalytic mechanism. It seems clear that haem a and Cu_A cannot simply be given an "electron wire" function in the enzyme. The kinetic anomalies may be most easily explained by the assumption that cytochrome a (and Cu_A) shuttles between two states during turnover.

7

Electron transfer and energy transduction

It has long been known that electron transfer between cytochrome c and dioxygen is linked to oxidative phosphorylation (Maley and Lardy, 1954; Lehninger et al., 1954). Traditionally, the cytochrome oxidase reaction has been considered to include the third "site", or simply "site 3" of oxidative phosphorylation. It was the pioneering studies of Racker, Hinkle and their collaborators (see Chapter 2) with isolated respiratory chain complexes re-incorporated into liposome membranes that established that energy conservation is a property of these complexes themselves. Thus Hinkle et al. (1972) and Hinkle (1973) showed evidence that cytochrome oxidase is capable of generating $\Delta\tilde{\mu}_{H^+}$ across such liposomal membranes. However, the quantitative aspects of these studies indicated catalysis of electron translocation and uptake of the "substrate" proton from the M phase in agreement with Mitchell's model (Chapter 2). This was presumably due to insufficient buffering of the intravesicular space in these vesicles which prevented detection of true proton translocation (Krab and Wikström, 1978; Wikström and Krab, 1979a).

There is no doubt that the chemiosmotic theory with its adhering experimentation has meant an important breakthrough in our understanding of the mechanism of oxidative phosphorylation. It is, however, a common but unfortunate misconception to believe that the chemiosmotic theory has solved the problem. The mechanism of oxidative phosphorylation may be broken into three separate events, viz. generation, transmission and utilization of proton currents (see Boyer et al., 1977; Wikström and Krab, 1980; Wikström et al., 1981). The molecular mechanism of none of these events has been elucidated, all three being the subjects of intensive research. With respect to the generation of $\Delta\tilde{\mu}_{H^+}$ the data on proton translocation by cytochrome oxidase (Chapter 2) are most relevant, because they show that at least for this respiratory chain complex the mechanism must differ from that proposed in the chemiosmotic theory. As will be demonstrated below, this difference is fundamental structurally, thermodynamically (see Chapter 2) and mechanistically. This may have contributed to the fact that proton translocation by cytochrome oxidase is not yet generally accepted.

Proton translocation catalysed by a formal electron-transferring enzyme complex is first analysed below on a general level which leads to development of a basic model of general applicability. This model yields certain precise predictions that may be tested experimentally for the case of cytochrome oxidase. We attempt to show in this chapter that analysis of available experimental data in the light of this model may lead to some true progress in our understanding of the electron transfer and proton translocation mechanisms of cytochrome oxidase. This chapter will end on a more speculative note where the results of the analysis have been used to construct a more specific, tentative model of cytochrome oxidase catalysis that is based on much of the information on structure, thermodynamics and kinetics discussed in previous chapters. While this model is more rational in biochemical terms than is our original, more general scheme, and therefore also better amenable to experimental test in the future, it is simultaneously also more likely to be in error. Only future work will tell whether this error is large or, as we would hope, comparatively small.

I. Molecular aspects of redox-linked proton translocation

A. General principles

The general principles of redox-linked proton translocation have been recently reviewed (Wikström and Krab, 1978, 1979a, 1980; Wikström, 1981a). One concept, that of the so-called redox loop, depends on group translocation of reducing equivalents, i.e. on vectorial organization of the oxidoreduction reactions (see Mitchell, 1979). The second "proton pump" concept is more general in that it accommodates translocation of the proton on residues that need not be the reduced forms of classical hydrogen transfer carriers. Proton translocation would thus be possible in a protein (such as cytochrome oxidase) that contains only formal electron carriers. In such a case the proton/electron coupling would necessarily be less "direct" (see also Mitchell, 1977) than in the redox loop case, and may even extend between different polypeptide chains, although more "directly" coupled cases could also be envisaged (see Wikström and Krab, 1978, 1979a). In a proton pump the transfer of reducing equivalents need not contribute to the electro-osmotic work of H^+ translocation. Further principles are outlined in more detail below. Here it suffices to state that the discovery of proton translocation by cytochrome oxidase (Chapter 2) has made it necessary to consider redox-linked proton pump types of mechanisms in oxidative phosphorylation.

B. Basic elements and properties of a redox-linked proton pump

Any proton-translocation mechanism requires a minimum of four elementary steps that take place cyclically, i.e. (i) uptake of H^+ by an acid/base group in protonic contact with the aqueous phase on one (M) side of the membrane; (ii) reorientation of the protonated group to render it in protonic contact with the other (C) side of the membrane; (iii) release of the proton into the C phase; and (iv) reorientation of the unprotonated group to re-establish contact with the M side.

Clearly the actual mechanism may be much more complicated. However, this basic sequence aptly demonstrates that two quite different kinds of molecular event are necessary, viz. protonation/deprotonation of one (or several) acid/base group(s), and "reorientation" steps in which protonic contact is established between acid/base groups alternately to the two aqueous phases on opposite sides of the membrane. When viewed in this way the cycle described seems trivial and obvious. Yet, the necessity of the "reorientation" steps, whatever their exact molecular meaning may be, is often neglected.

It is equally obvious that the above elementary cycle must be linked to the redox reaction. In principle, this linkage could be established to any one of the four elementary steps. However, in cytochrome oxidase, as in other redox-linked proton pumps that may exist, it is very likely that the link is established to the protonation/deprotonation steps. This is indicated by the pH dependence of the E_m values of redox centres (Chapter 5), showing directly that the redox event is coupled to shifts in pK of acid/base groups in the enzyme complex.

In summary, a redox-linked proton pump consists of a *redox centre* that catalyses the exergonic transfer of reducing equivalents, the redox state of which is coupled to the function of one or several *acid/base groups*. Moreover, one or more such groups must have the ability for "*reorientation*", by which protonic contact is established in a controlled fashion alternately between the two aqueous phases.

C. A general model of redox-linked proton translocation

After recognition of the basic elements and reaction types of a proton pump it may be useful to design a functional model that could provide the framework for further discussion and, above all, for experimental testing. In the initial stage it is essential that the model is general enough not to limit progress and yet simple enough to allow meaningful experimental tests.

Figure 7.1 depicts a model proposed by Wikström and Krab (1979*a*) which satisfies these criteria. In this model the three types of reactions, viz. oxidoreduction, protonation/deprotonation and reorientation, are separated along the three Cartesian coordinates to yield a cubic scheme consisting of eight different states. In many respects this is a minimal model. For

Low potential ^{M}OX ← OX^C

e^- →

^{M}RED - - -|- - - - - - RED^C → H^+_C

H^+_M → H^{+M}OX - - - - - - -|- - - OX^{C}H$^+$

→ e^-

High potential

H^{+M}RED → RED^{C}H$^+$

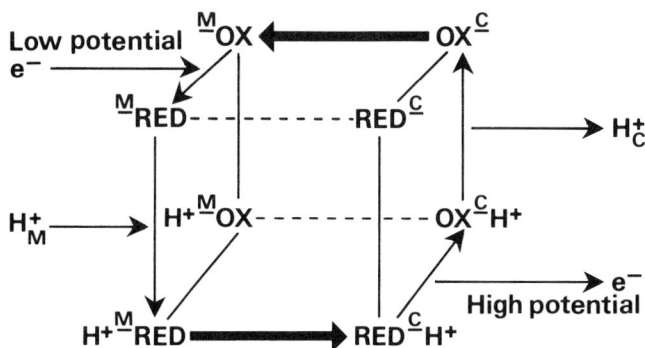

Fig. 7.1 Cubic scheme of a redox-linked proton pump. RED and OX denote reduced and oxidized states of the redox centre. Bars with M or C denote a redox-linked acid/base group in protonic contact with the aqueous M and C phases. respectively. Both protonated and unprotonated forms of this group are shown. Thick arrows show reorientation steps in which (i) the redox centre is switched between electron input and output modes (left and right, respectively), and (ii) the acid/base group is switched simultaneously between protonic contact with the M (input) and C (output) phases. Transitions along broken lines would lead to futile cycles (molecular "slipping"), in which electrons may move across the transducer without coupling to proton translocation, or vice versa. Thin plus thick arrows depict one of several possible reaction paths. *Note*: Transitions along the x-axis denote "reorientation" with respect to electronic and protonic input/output; transitions along the y-axis denote protonation/deprotonation; transitions along the z-axis denote oxidoreduction. Modified from Wikström and Krab (1979)

example, only one acid/base group has been implicated. Moreover, the input states of the electron and the proton have been merged together, as have also the respective output states. The latter simplifications are required to keep the model three-dimensional. However, they may indeed be appropriate for cytochrome oxidase on the basis of experimental evidence (Section II.C). It follows from these assumptions that the reorientation steps take place simultaneously for both the electron and the proton transfer reactions.

The model demonstrates that some reorientational transitions (dotted lines, Fig. 7.1) should have much lower probability (rate constants) than others to prevent short-circuiting of the transducer (molecular "slipping").

As has been pointed out by De Vault (1971), it is necessary to separate input and output states of the redox element in a redox energy transducer. For analogous reasons it is also necessary to separate the input and output states of the acid/base group. Any simplification in these respects leads to significant loss of detail and a decrease in the predictive power of the model.

With the above provisions the model in Fig. 7.1 is completely general and applies to any redox-linked proton translocator. Moreover, it can easily be modified to describe any energy transducer (e.g. the proton-translocating ATPase). Note especially that the molecular nature of the reorientation steps has not been specified, but that the indicated arrowed pathway is a purely formal description.

It may be of interest to point out that the model in Fig. 7.1 also describes fully the proton/electron symporter unit of a redox loop. This comparison may be useful because it brings forward a facet of the proton pump that may not be immediately apparent from the scheme. Figure 7.1 describes an electroneutral H^+/e^- symport if (but only if) the path of the electron and that of the proton follow the same profile of electrostatic potential. In a proton pump, however, it is generally assumed that the path of the two particles across the transducer is completely different (cf. indirect coupling). Thus, in the simplest case the path of the electron is indifferent with respect to the electrostatic gradient across the membrane, so that only the flow of protons performs electrical work. However, intermediate cases may also be envisaged in which part of the electrical work is carried out by electron translocation and part by proton translocation.

Assuming, for simplicity, that only proton transfer performs electrical work, this work may be carried out by the reorientation steps (Fig. 7.1), either by transferring a positively charged protonated acid/base group (histidine?) from contact with the M phase into contact with the C phase or, alternatively, by switching a negatively charged unprotonated group (carboxylate?) in the opposite direction. The amount of electrical work performed in either of these steps depends entirely on the electrical potential difference across the transducer. At one extreme, this difference might approach zero if the transducer is equipped with "deep enough" proton wells, rendering the input and output states in protonic contact with the M (left) and C(right) aqueous phases, respectively. At such a limit the entire $\Delta\bar{\mu}_{H^+}$ *across the transducer* may be in the form of a concentration difference of protons (ΔpH; cf. Mitchell, 1968, for discussion of the concept of a proton well).

The reorientation steps (Fig. 7.1) do not imply any large movements in molecular terms. Controlled gating of proton channels may well suffice to render the acid/base group alternately in contact with the M and C phases. Another possibility would be a subtle movement of the acid/base group between proton-conducting channels leading to opposite sides of the membrane. The corresponding switch of the redox element between input and output states may also involve only a small conformational change, such as a shift in the geometry of the redox centre itself or of its immediate environment in the protein.

D. Shifts in E_m and pK and kinetic performance of the pump

Redox-linked shifts in the pK of acid/base groups (so-called redox Bohr effects) are often implicated in discussions of proton pump mechanisms (cf. Papa, 1976). However, as may be deduced from Fig. 7.1 (and see Wikström and Krab, 1979a), such shifts are not absolutely required for the translocational mechanisms *per se* (cf. Boyer, 1975). Yet, redox state-dependent pK shifts are indeed expected in this type of mechanism, but mainly for reasons of kinetic versatility and energetic efficiency.

If we assume that the pump exhibits high specificity so that "slipping" (dotted lines, Fig. 7.1) transitions have insignificant probability, it follows that the intrinsic H^+/e^- stoicheiometry of proton translocation will be unity and completely unaffected by changes in pH or electrical potential. Such specificity is by no means unheard of in enzymic systems, a well known example being the obligatory coupling between oxidoreduction and phosphorylation in the glyceraldehyde phosphate dehydrogenase reaction. Yet, we may also foresee conditions under which the probability of the "forbidden" transitions is raised significantly (Sections II.D, II.E and IV.E), either due to experimental perturbance or, perhaps, as a desired control function.

Generation of pH and electrical gradients by the pump would soon tend to limit it kinetically. This may be a desired effect (respiratory control), but should not set in until a considerable $\Delta\tilde{\mu}_{H^+}$ has been built up across the membrane. This is one reason why it may be desirable that the pK of the acid/base group be modulated according to the redox state of the redox centre. The observed modulation (Chapter 5) favours protonation of the reduced and deprotonation of the oxidized forms of the transducer, which may have several favourable effects. First, it tends to lower the probabilities of the "forbidden" transitions (cf. above) that lead to uncoupling ("slipping") of the pump. Provided that these transitions have low probabilities, this modulation would, secondly, tend to delay kinetic blockage to higher values of $\Delta\tilde{\mu}_{H^+}$ than would be the case without the modulation. Finally, the low $\tilde{\mu}_{H^+}$ at the input and the high $\tilde{\mu}_{H^+}$ at the output tend to adjust the respective E_m values of the input and output redox couples low and high. In this way the E_m values may be modulated to give optimal matches to the input and output redox potentials (E_h). If such potential matching did not occur, either reduced or oxidized forms of the transducer would predominate dramatically and prevent smooth overall kinetic performance (cf. De Vault, 1971).

The above modulations may be sufficient and shifts in pK between the input and output states need not be invoked in a minimum model. We should add, finally, that a proton pump may require further linkage phenomena in addition to those mentioned to perform well kinetically.

Assume, for example, that the transition from the reduced protonated input state to the corresponding output state (see Fig. 7.1) involves translocation of positive electrical charge and represents a step at which significant electrical work is done. This transition may then become considerably decelerated in the forward direction, even at relatively low electrical potential difference across the membrane. Such a situation may require linkage of this transition to another reaction step in the cycle that may be sufficiently exergonic to facilitate the kinetics of charge translocation. In Section IV of this chapter we propose a mechanism in which such linkage occurs between the forward and backward elementary reorientation steps shown by thick arrows in Fig. 7.1.

E. Proton-conducting channels

A proton translocator is expected to be equipped with specific proton-conducting pathways, e.g. to shorten distances in possible "reorientation" mechanisms. These proton channels may have an entirely passive proton-conducting function, or may be involved with any degree of complexity with the H^+-translocating mechanism itself. One possible more complex function would be to provide controlled gating of the proton conductances (cf. Boyer, 1980). In Fig. 7.1 this would imply control or catalysis of the reorientation steps.

Proposed proton-conducting structures in proteins are based on hydrogen-bonded structures along which protons may "move" by quantum-mechanical tunnelling (Morowitz, 1978; Nagle and Morowitz, 1978; Kayalar, 1979; for a review see Wikström et al., 1981).

F. Experimental predictions from the model

A theoretical scheme such as that of Fig. 7.1 is of limited value unless it may be tested experimentally. In this sense it turns out that Fig. 7.1 is, in fact, quite powerful. In this section we make note of some basic predictions from Fig. 7.1, which have important implications for the cytochrome oxidase proton pump, and which will be tested directly in the sections to follow.

The redox element of the proton pump is expected to exhibit several unique properties. First, it seems likely that the E_m of such a centre should be a function of pH. Second, the centre should be able to exist in two states, in which it is in electronic contact with the electron donor and acceptor sides of the transducer, respectively. Our specific model predicts further that the electronic input state of the redox centre may coincide with the input state of the acid/base group, and that the electronic output state coincides with the output state of the acid/base group. Thus, if the input and output states of the redox centre could be studied separately, their respective E_m values should be functions of pH uniquely in the M and C phases, respectively.

Third, it is conceivable that the "pure" electron transfer steps of electron input and output may be faster and less temperature-sensitive than the "reorientations" of the redox centre, due to the conformational nature of the latter. If so, the two states might be distinguished by kinetic experiments, particularly at low temperatures.

Finally, it may be worth emphasizing that we expect the basic properties of the proton pump to be retained in the cytochrome oxidase molecule as isolated, because the proton pump has been shown to be inherent in the oxidase itself by reconstitution experiments (Wikström and Saari, 1977; Krab and Wikström, 1978; and see Wikström and Krab, 1979a).

II. The proton pump of cytochrome oxidase

A. Identification of the redox element

The haems of cytochrome oxidase are much more likely candidates than the coppers as redox elements of the proton pump due to the pH dependence of their midpoint potentials (Chapter 5), which is not shared by the coppers. However, a *prima facie* choice between the two haems in this respect may be more difficult.

A priori, cytochrome a seems a better candidate than a_3 since the latter is intimately involved in the O_2-reducing function (Chapter 6). Haem a_3 and its associated copper must cycle through at least four different states in the reduction of an O_2 molecule. It is therefore hard to envisage how this cytochrome could be more directly involved in proton translocation in the sense shown in Fig. 7.1.

Apart from some early proposals of an involvement of cytochrome a in energy transduction (Wilson and Chance, 1966, 1967; and see also Wikström and Krab, 1979a), it is cytochrome a_3 that has more recently been primarily implicated in such a function. The reason for this is the large, energy-dependent spectral shift in this cytochrome, observed in mitochondria, and suggested to correspond to a transition of the cytochrome into a "high energy" state directly involved in energy transduction (Erecińska et al., 1972; Wilson et al., 1973) or, indeed, in proton transloca-tion more specifically (Wikström, 1975, 1977; Wikström and Saari, 1977; Wikström and Krab, 1978, 1979a).

However, one problem associated with these proposals has all along been the general question of why only cytochrome a_3 should exhibit such a spectrally detectable "high energy" form, whereas no analogous species are observed at other energy-transducing sites in the respiratory chain. Although this question does not contradict the above proposals, it is nevertheless disturbing. It was therefore rewarding that a natural explanation to this problem could recently be provided (Wikström, 1981c). In this work (and see Section III.A for a more detailed description) it was shown that the energy-dependent spectral shift is simply the result of reversed electron transfer from water to cytochrome c, in which process the fully oxidized state of the haem a_3/Cu$_B$ centre reverts stepwise to two earlier states in the catalytic cycle of O_2 reduction, viz. the states possibly corresponding to nos **2** and **1** respectively in Fig. 6.6. The spectral shifts associated with this process thus correspond closely to those expected on transition of the oxidized centre to intermediate states identified in the forward reaction of the reduced enzyme with O_2.

This finding provides a self-consistent explanation for the fact that cytochrome a_3 is the only component of the respiratory chain that exhibits a large spectral shift upon "energization". It is the only redox centre in which a dramatic change in the ligand state must accompany the elementary electron transfer steps. This also removes the basis for the proposals that cytochrome a_3 would necessarily be directly involved in proton translocation. Since our *a priori* considerations favour cytochrome a over a_3 in such a role (see above), we may now proceed to test whether cytochrome a shows any properties that may be ascribed to such a function.

B. The two states of cytochrome a

It was predicted (Section I.F) that the redox element of the proton pump should exist in two interconvertible states of electron input and output, respectively, that may be kinetically distinguishable if the "reorientation" or interconversion between these states is rate-limiting for overall electron transfer. This prediction is uniquely fulfilled by cytochrome a, as shown by three different and independent kinetic tests.

It was shown in Chapter 6 that the kinetics of oxidation of cytochrome a is heterogeneous following a pulse of O_2 to the reduced enzyme. Thus about 30–50% of the cytochrome appears to be rapidly oxidized immediately following oxidation of the haem a_3/Cu$_B$ centre, whereas the rest is oxidized in a much slower step in conjunction with oxidation of cytochrome c. This distinction is indeed much enhanced at low temperatures (Erecińska and Chance, 1972), and at temperatures below $c. -78 \,°C$ only the first phase occurs (Compound B; see Chapter 6). It may be noted that this kinetic heterogeneity is also "mirrored" in the behaviour of Cu$_A$, which appears to follow cytochrome a kinetically. We conclude that if cytochrome a were the redox element of the proton pump this kinetic heterogeneity would find a rational explanation.

If our interpretation is correct, the kinetic distinction should also be possible in reductant-pulse experiments, provided that electron input is sufficiently fast to render the "reorientation" step rate-limiting. As discussed in Chapter 6, the fast phase of reduction of the enzyme by ferrocytochrome c indeed comprises only one-half of haem a (and one half of Cu_A), no matter how high the applied c^{2+}/aa_3 ratio. This agrees precisely with our prediction. These data thus indicate that it is the electron input mode of cytochrome a that is reduced in the fast phase, that Cu_A "follows" haem a in its state transitions (cf. above for the O_2 pulse), and that the transition between input and output mode of cytochrome a may be particularly slow in the fully oxidized "resting" enzyme.

Both the O_2- and cytochrome c-pulse experiments indicate that input and output states of cytochrome a may be about equally "occupied" at all times. This may be an important piece of information, the possible significance of which is discussed below (Section IV).

If a kinetic distinction is possible between input and output electronic states of the transducer in O_2- and reductant-pulse experiments, it may also be possible in the steady state. In the case in which electron input from cytochrome c is sufficiently fast the "reorientation" steps may become rate-limiting for overall activity. In such a case, any further increase in the electron input rate should increase the steady state level of reduction of the input state of cytochrome a alone, whereas the output state should remain highly oxidized due to fast electron transfer to O_2 via the a_3/Cu_B centre.

Yonetani (1960b) showed that when cytochrome oxidase is titrated aerobically with ascorbate in the presence of cytochrome c, the latter becomes linearly more reduced, until a steady state degree of reduction near 100% is reached at high ascorbate concentrations. In contrast, the haem absorption at 605 and 445 nm increases to maxima of only 60 and 30% of the absorption in the fully reduced enzyme. No further increase in absorption is possible by increasing the ascorbate concentration further in the aerobic steady state. This result would be explained simply as due to reduction of haem a but not of a_3 in the aerobic steady state were it not for the fact that the spectral contribution of haem a is much larger than 60 and 30%, respectively, at the two wavelengths (Chapter 4). Therefore this finding again indicates a "heterogeneity" in cytochrome a of precisely the kind expected from Fig. 7.1. It is noteworthy that the "occupancy" of input and output states of cytochrome a again appears to be roughly 50/50 as in the O_2- and reductant-pulse experiments.

We conclude that cytochrome a shows a unique kinetic heterogeneity in three different and independent tests. This heterogeneity is of precisely the kind expected from a redox centre of an energy transducer (Section I.F, Fig. 7.1). It is therefore suggested that cytochrome a is the redox element of the proton pump in cytochrome oxidase.

C. Sidedness of the redox-linked acid/base groups

In Section I.E it was predicted that the input state of the redox element should exhibit an E_m which is uniquely a function of pH in the M phase (see also Fig. 7.1). Conversely, the model predicts that the E_m of the "output" couple should depend on pH in the C phase.

Artzatbanov et al. (1978) showed that the redox equilibrium between cytochromes c and a is specifically dependent of pH in the M phase in mitochondria when electron transfer between cytochrome a and the a_3/Cu_B centre is blocked by cyanide. We have subsequently confirmed this finding (see Wikström and Krab, 1979c; Wikström, 1981b). Since the E_m of cytochrome c is independent of pH near neutrality (Rodkey and Ball, 1950), the observed pH dependence can be entirely attributed to cytochrome a. Therefore this finding is in remarkable agreement with the predictions of Fig. 7.1, strongly supporting our conclusion that cytochrome a is the redox element of the proton pump. This finding specifically supports the simplification in the model according to which input and output states of the electron and the proton have been merged together. Further experimentation along these lines may provide important insights into the mechanism of proton translocation.

Reversed electron transfer, as observed by the energy-dependent spectral shift in ferricytochrome a_3 (Wikström, 1981c), is associated with proton uptake from the C phase (Wikström, 1975). This also supports, albeit indirectly, the converse prediction that the redox-linked acid/base group is in protonic contact with the C phase when cytochrome a is in the electronic output state.

D. A possible role of subunit III in proton translocation

It was suggested previously (Wikström and Saari, 1977; Wikström and Krab, 1979a), by analogy with the function of the F_0 segment of the mitochondrial H^+-translocating ATPase (cf. Kozlov and Skulachev, 1977), that one of the mitochondrially synthesized subunits of cytochrome oxidase (Chapter 3) might furnish the proton pump with a H^+-conducting channel. This notion has recently gained experimental support.

Reconstitution of cytochrome oxidase devoid of subunit III (Penttilä et al., 1979) into phospholipid vesicles yields a preparation that is active with respect to overall redox activity and has intact general optical spectra, but lacks measurable proton translocation. Yet, these subunit III-free preparations show excellent respiratory control (Saraste et al., 1981). Although reconstitution of subunit III back to the enzyme has not yet been achieved, these data suggest that subunit III may have an important role in proton translocation. The lack of the proton pump in the subunit III-free preparation was recently confirmed by the finding that this preparation exhibits a K^+/e^- ratio of potassium uptake (+valinomycin) near unity (Penttilä and Wikström, 1981), whereas this ratio is near 2 with the intact enzyme (Sigel and Carafoli, 1979, 1980; Coin and Hinkle, 1979).

These findings are reminiscent of the recent work with cytochrome oxidase isolated from *Paracoccus*. In the isolated state this enzyme contains only two different subunits, which resemble subunits I and II of the mitochondrial enzyme (Ludwig and Schatz, 1980; Ludwig, 1981). After reconstitution into liposomes, good respiratory control is observed, but little or no proton translocation. Yet, it was shown recently that the *Paracoccus* enzyme is indeed a proton pump in the bacterial membrane (Van Verseveld *et al.*, 1981). It is thus possible that a putative subunit, corresponding to subunit III in the mitochondrial enzyme, is lost on isolation of the *Paracoccus* cytochrome oxidase. This would then be reminiscent of the relative ease by which subunit III may be removed from the beef heart enzyme (Penttilä *et al.*, 1979; Saraste *et al.*, 1980, 1981).

Casey *et al.* (1980) suggested that dicyclohexylcarbodiimide (DCCD; cf. Beechey *et al.*, 1967; Cattell *et al.*, 1971) blocks proton translocation in cytochrome oxidase through specific binding of this carboxyl group reagent to subunit III. Binding of DCCD to a glutamyl residue in subunit III was subsequently demonstrated by Prochaska *et al.* (1981). Hence there seems to be a striking analogy to the inhibition of the H^+ channel of H^+-translocating ATPases by DCCD (Sebald *et al.*, 1980), supporting the notion that subunit III may function as such a channel. This analogy is further supported by the fact that the glutamyl residue seems to be in the middle of an otherwise very hydrophobic sequence in both proteins (cf. Chapter 3 and see discussion by Wikström *et al.*, 1981). The primary structure of subunit III suggests that it may form seven α-helices that traverse the membrane (see Chapter 3). Azzi (1980) has drawn attention to the apparent analogy between this and the membranous structure of the proton-translocating bacteriorhodopsin.

The lack of respiratory inhibition and retainment of respiratory control in liposomes containing the subunit III-free enzyme indicates that subunit III may play a more active role in proton translocation than merely forming a passive H^+-conductance. A simple blockage of a proton channel should lead to simultaneous inhibition of electron transfer provided that the two processes are tightly coupled. On the other hand, the retained respiratory control indicates that energy conservation still occurs, and that the membranes have not been rendered permeable to protons. These data suggest then that the proton pump is locally uncoupled, which is equivalent to molecular "slipping" of the coupling process (see Section I.C). Hence subunit III may be responsible for a gating mechanism that co-ordinates "allowed" reorientation steps of the pump (Fig. 7.1) with oxidoreduction. In the absence of this polypeptide the "forbidden" reorientations (dotted lines; Fig. 7.1) may acquire high probability relative to the normal coupled route. The reason why respiratory control is yet retained is discussed in the next section.

Subunit III may, in fact, be even more intimately involved with the H^+ pump mechanism than anticipated from the above evidence. Penttilä and Wikström (1981) recently demonstrated that the pH dependence of the E_m of cytochrome a is abolished in the pH 6.5–8 region by removal of subunit III. This finding has far-reaching corollaries. First, it lends strong support to the idea that cytochrome a and its redox-linked acid/base group are intimately involved in the proton pump mechanism. Second, it suggests that this acid/base group might be located on subunit III, although this is not proved by the available data. In order to test whether the DCCD-sensitive glutamyl residue on subunit III (Prochaska et al., 1981) might be the redox-linked acid/base group, we tested the effect of DCCD on the "seven subunit" enzyme and on intact mitochondria (T. Penttilä and M. Wikström, unpublished). The results were ambiguous, however, DCCD sometimes causing a change in the pH dependence of the E_m. Moreover, in most cases this change was very small, particularly in intact mitochondria. More work is required on this matter. On the other hand, the nature of the change in the pH dependence of cytochrome a's midpoint potential on removing subunit III seems to be in better accordance with an acid/base group that is charged in the protonated state. A histidine residue is therefore a distinct possibility (Penttilä and Wikström, 1981). Further work with group-specific modifying reagents may provide an answer to this question.

In conclusion, several independent lines of evidence suggest that subunit III has a specific and important role in the proton-translocating mechanism of cytochrome oxidase. The data are consistent with the possibility that the essential H^+–e^- coupling takes place between cytochrome a (subunit II?) and subunit III, i.e. between different polypeptide chains.

E. The role of the haem a_3/Cu_B centre in energy transduction

The finding that respiratory control is retained in a reconstituted oxidase preparation devoid of subunit III, although the H^+ pump appears to be intrinsically uncoupled (cf. above), provides important information on the contribution of the haem a_3/Cu_B centre to energy conservation.

As concluded earlier (Wikström, 1977, 1981d), there was previously no way to distinguish from the available data whether (i) the proton pump catalyses translocation of $2H^+/e^-$ with uptake of the "substrate proton" (cf.

Chapter 2) from the C side of the membrane, or (ii) whether the pump translocates $1H^+/e^-$, in which case the "substrate proton" must be taken from the M side. Both alternatives yield the overall release of $1H^+/e^-$ on the C side, uptake of $2H^+/e^-$ from the M side and translocation of two electrical charge equivalents per transferred electron (cf. Fig. 2.4). Both alternatives are also energetically equivalent, corresponding thermodynamically to translocation of $2H^+/e^-$ (cf. Wikström *et al.*, 1981). They are, however, mechanistically very different indeed, because in (i) the proton pump is responsible for the entire energy conservation by the enzyme, whereas in (ii) half of the overall energy conservation is due to the H^+ pump and the other half is due to the function of the a_3/Cu_B centre in accepting electrons from the C side and H^+ from the M side of the membrane.

The finding that local uncoupling of the H^+ pump does not abolish respiratory control and leaves half of the electrical charge translocation of the enzyme is compatible only with the second alternative (ii). It therefore seems that the subunit III-free enzyme functions in a way analogous to the function suggested by Mitchell (see, for example, Mitchell and Moyle, 1967) for the native enzyme. In the intact enzyme, therefore, the proton pump is apparently connected in series with the charge-separating function of the haem a_3/Cu_B centre (Penttilä and Wikström, 1981; Wikström, 1981*e*). If this proposal is correct, it may have interesting evolutionary implications (see Epilogue).

III. Reversed electron flow and other energy-dependent phenomena

A. Partial reversal of the O_2 reaction

It is well known that electron transfer may be reversed in the respiratory chain at a high $ATP/ADP \cdot P_i$ ratio (see Klingenberg and Schollmeyer, 1961). The only step that has traditionally been considered irreversible is the final reduction of O_2 to water by the haem a_3/Cu_B centre. Exhaustive attempts to reverse this reaction have consistently yielded negative results, or possibly extremely minute amounts of O_2 produced from water (see Bienfait, 1975).

However, the irreversibility of the O_2 reaction *in toto* does not exclude the possibility that some partial steps (cf. Fig. 6.6) might be reversible so that ATP-dependent "reversed electron transfer" might generate states of the haem a_3/Cu_B centre that are intermediates in the forward reaction with dioxygen. If this occurred, one would expect to observe sizable spectral shifts specifically in haem a_3 generated under the appropriate conditions favouring reversed electron transfer.

It was indeed shown recently (Wikström, 1981c) that the previously observed energy-dependent spectral shift in ferric haem a_3 (Erecińska *et al*., 1972) can be attributed to such reversed electron transfer. Titrations with oxidoreduction potential revealed that the compound described by Erecińska and her colleagues is formed from the fully oxidized enzyme coupled to transfer of one electron from water to cytochrome c. This intermediate was therefore ascribed to be the ferryl haem iron–Cu_B^{II} species (no. 2 in Fig. 6.6), and indeed has an optical spectrum resembling the "Compound II" of Orii and King (1972, 1976), and the $g = 5$ species of Shaw *et al*. (1978; cf. Chapters 4 and 6).

It was also found that if the ATP/ADP·P_i ratio, or the so-called phosphorylation potential, was further raised, or if the E_h was raised at cytochrome c, Erecińska's compound was replaced by a species with spectral properties identical to those of Compound C (Chapter 6). Since this transition was also found to be a one-electron step (Wikström, 1981c), this result conforms well to the proposal (Chapter 6) that Compound C is a peroxidic intermediate.

No further reversal of the oxygen reaction was achieved at phosphorylation and redox potentials managable with ATP and ferricyanide, respectively.

These findings lead to several important corollaries. First, they remove the previous experimental indication for a direct role of cytochrome a_3 in energy transduction, and provide a rational explanation for the fact that haem a_3 is unique in exhibiting a large energy-dependent spectral shift in mitochondria (cf. Section II.A). Indirectly this provides more strength to our conclusion (cf. above) that cytochrome a may be involved directly in the function of the proton pump. Second, these data provide strong support for the contention that Compound C is a true intermediate in O_2 reduction (cf. Chapter 6), and that it cannot be an artefact due to CO (Nicholls, 1978, 1979a,b). Third, they support the idea that Compound C is a peroxy species. Due to the very oxidizing conditions during reversed electron transfer (see Wikström, 1981c), it is very unlikely that haem a_3 would remain ferrous in this compound (cf. Chapter 6). Fourth, the results suggest that the different steps of O_2 reduction at the a_3/Cu_B centre may be energetically different. Thus the irreversibility of the overall reaction could be due to a more local irreversibility of the step in which bound O_2 is reduced to bound peroxide (see Chapter 6), whereas other steps may be more or less easily reversible at sufficiently high redox and phosphorylation potentials. We should point out here, however, that the conditions in which partial reversal of the O_2 reaction was observed were such that reversal would hardly occur to a significant extent under physiological conditions. Even though a high phosphorylation potential may be achieved in the so-called resting state of mitochondria (State 4; see Chance and Williams, 1955), it is unlikely that this would be associated with a redox state of cytochrome c corresponding to a redox potential much above 300 mV. At this limit there would be approximately 10% conversion of the fully oxidized form into the ferryl iron–Cu_B^{II} species at equilibrium (cf. Wikström, 1981c).

Last, but not least, these data provide a new and independent method of studying the coupling between electron transfer and proton translocation by cytochrome oxidase. Preliminary experiments (M. Wikström, unpublished data) have suggested that each of the one-electron steps in the backward reaction is associated with translocation of two electrical charges across the mitochondrial membrane. If this can be verified, it would provide further strong support to the notion (Wikström, 1977; Wikström and Saari, 1977; Krab and Wikström, 1978) that cytochrome oxidase translocates two electrical charges per transferred electron (see also Chapter 2).

B. The spectral shift in ferrocytochrome a

Ferrocytochrome a is slightly shifted spectrally under "high energy" conditions of isolated mitochondria (Wikström and Saris, 1970; Wikström, 1972). This spectral change can be mimicked by Ca^{2+} and H^+ ions in uncoupled mitochondria and in the isolated enzyme (Wikström and Saari, 1975; Saari et $al.$, 1980). Since the energy-dependent shift does not appear to be caused by Ca^{2+}, but may be mimicked by this cation from the C-side of the membrane, it was suggested that the energy-dependent effect might be due to a local increase of H^+ activity at the binding site on the oxidase.

More recent data (Saari et $al.$, 1980) showed that a similar spectral change may be induced by cations with isolated haem A. This effect was abolished by esterification of the propionate side chains of the haem. Together these data then indicate that in the "energized" state of intact mitochondria the vicinity of the haem of cytochrome a may become considerably more acidic, notably in the region near the propionate side chains. No such effect is observed for cytochrome a_3 (Wikström and Saari, 1975), even though haem a_3 also reacts with Ca^{2+} in a similar fashion in the isolated enzyme. However, in the membranous enzyme haem a_3 appears to be shielded from Ca^{2+} (Saari et $al.$, 1980).

These data provide some indications of the location of haems a and a_3 in the membranous enzyme (see Chapter 3). Moreover, they might be related to our proposal that cytochrome a is primarily involved in the functioning of the proton pump. It is not excluded that structures on haem a might be directly involved in proton translocation, possibly in conjunction with residues on the subunit III polypeptide (Section III.D). Possible candidates for the former type of residues are then the propionate carboxyls (Saari *et al.*, 1980) and the carbonyl group (Babcock *et al.*, 1981); see also Chapter 4). If the latter is involved, an interesting analogy might arise between the proton pumps of cytochrome oxidase and bacteriorhodopsin. In the latter system a Schiff's base involving a carbonyl group of the chromophore has been directly implicated in the mechanism of proton translocation (see, for example, Lewis *et al.*, 1978).

C. Energy-dependent changes in measured E_m values

Wilson and his collaborators (Wilson and Dutton, 1970; Wilson *et al.*, 1972*b*) showed that the measured E_m values of the haem system (see Chapter 5) were changed considerably under anaerobic conditions when mitochondria were "energized" by ATP. If analysed according to the neo-classical model, these changes appear to consist of an ATP-induced decrease in the measured E_m of both cytochromes a and a_3 (Wikström *et al.*, 1976).

An explicit interpretation of these changes is difficult (see, for example, Walz, 1979). It may be noted, however, that these effects appear to be confined to the haems alone. Neither Cu_A nor Cu_B exhibits any significant E_m shift (Erecińska *et al.*, 1971; Lindsay *et al.*, 1975), although the latter can be measured only indirectly. The main obstacle for a straightforward interpretation of this type of data is the uncertainty of the point of interaction of the redox mediators used with the measured system. Mediators employed in this type of experiment are usually known to interact with mitochondria through cytochrome c. If this is the sole point of redox interaction, the phenomenon of reversed electron transfer will tend to cause an apparent lowering of the E_m of all redox couples that equilibrate with cytochrome c via a "site" of energy coupling (see De Vault, 1971). However, as also pointed out by Lindsay *et al.* (1975), the apparent lack of an effect on the E_m of Cu_B is difficult to reconcile with this comparatively simple explanation. Similarly, if the apparent E_m of haem a is indeed decreased, it is difficult to understand why there is not an associated decrease in the measured E_m of Cu_A, if the latter indeed equilibrates exclusively via haem a (Chapter 6). The interpretation of these phenomena is further complicated by the possibility of a "sidedness" of the effect of pH on the redox centres (cf. Section III.C and III.E). As pointed out by Wikström and Krab (1979*a*), the effect of "energization" on measured E_m values indeed correlates with the effect of pH, the haems—but apparently not the coppers—being affected in both cases. Further complications might arise from differential effects of ΔpH and $\Delta\psi$ on partial reactions in the oxidase complex and from the duality of cytochrome a (Section II.B). It seems clear that a complete study on the mode of action of different redox mediators is required first, but that more experimentation on this effect is also required more generally before it can be unambiguously interpreted (see note 1 on p. 190).

Hinkle and Mitchell (1970) measured the effect of an applied electrical potential difference across the membrane on the redox equilibrium between cytochromes c and a in CO-inhibited mitochondria. The E_h of cytochrome c was clamped by ferri-/ferrocyanide. They observed an apparent shift in E_m of haem a which was roughly one-half of the applied electrical potential. When the polarity of the latter was positive in the C-phase the E_m apparently decreased, and vice versa. This was interpreted to mean that electron transfer from cytochrome c to cytochrome a takes place across about one-half of the membrane dielectric. Mitchell and Moyle (1979) concluded more recently that these results exclude any energetic role of the redox changes of cytochrome a in a proton pump. However, this conclusion is premature for several reasons. First, it was not excluded that the observed effect might be due to a $\Delta\psi$-induced change in the H^+ activity in a proton channel or well (see Mitchell, 1968) connecting the redox-linked acid/base group to the M-phase (cf. Section II.C). Second, elementary reaction steps of a proton pump in which cytochrome a serves as the redox element (see Sections I and II) may include purely $\Delta\psi$-dependent transitions. Either of the two "reorientation" steps (Fig. 7.1) may be associated with effective translocation of electrical charge. The work done in such a step depends on the electrostatic potential profile across the transducer itself. This potential drop may be any fraction of the total $\Delta\psi$ between the aqueous phases. Finally, the interpretation of the phenomenon described by Hinkle and Mitchell (1970) is beset with many of the difficulties discussed above for energy-dependent E_m shifts more generally. The fact that the coppers do not seem to respond to "energization" (see above) is difficult to reconcile with the proposal that the effect of $\Delta\psi$ is a trans-spatial phenomenon, acting on electron transfer reactions orientated vectorially across the membrane (Wikström and Krab, 1979a).

We conclude that the data reported by Hinkle and Mitchell (1970) provide no unique indication for an electron-translocation function of cytochrome oxidase. They certainly do not contradict the notion of cytochrome oxidase as a proton pump, or of cytochrome a as an essential element thereof.

IV. A reciprocating site mechanism

In this section we will present a tentative model of electron transfer and proton translocation which represents an attempt to rationalize the available experimental information with the cubic model in Fig. 7.1. Since the

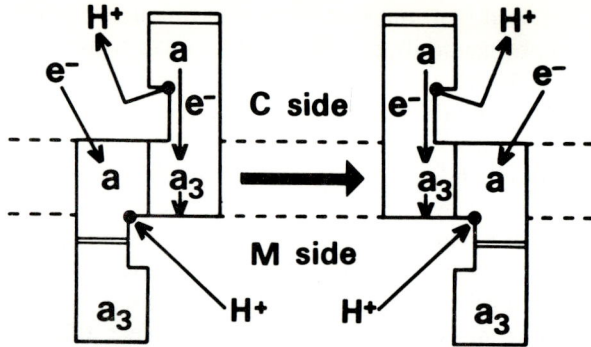

Fig. 7.2   Reciprocating site model of electron transfer and proton translocation by dimeric cytochrome oxidase. Two states of the dimers are shown on each side of the thick arrow. These differ from each other by a sliding movement of the monomers in opposite directions (upwards/downwards) along an axis perpendicular to the plane of the membrane. The two states are exactly equivalent, except that the monomers have acquired opposite roles of input (downward) and output (upward), respectively. In the input state electron transfer is possible beween haems of cytochromes c and a, but not between haem a and the a_3/Cu_B centre (as indicated by the double line). Conversely, in the output state electron transfer occurs between haem a and the a_3/Cu_B centre, but not between haem c and haem a (indicated again by the blocking double line). In the input state a redox-linked acid/base group (black dot near haem a) communicates protonically with the aqueous M phase. In the output state this group communicates with the C phase. Dotted lines symbolize an electro-osmotic barrier in the protein to illustrate the different orientations of the acid/base group. After completion of electron transfer and H^+ release or uptake in each monomer (left), there is a reciprocal switch (thick arrow) by which the monomer previously in the input state is converted into the output state and vice versa. After completion of catalytic steps on the right of the thick arrow an analogous switch brings the enzyme back to the situation on the left, etc. Note especially that this switch has been drawn large only for illustrative purposes.

latter model is abstract and rather general, it may be useful to see how it might be made more concrete. It is clear that such a step could increase the possibilities for experimental testing.

We were struck by the finding from all kinetic tests of cytochrome a (Section II.B) that the input and output states seemed to be equally occupied under all conditions. Although this might be coincidental, there is the intriguing possibility that this is due to co-existence of input and output states on the same catalytic unit, which would then be the $(aa_3)_2$ dimer. The results from the cytochrome c-pulse experiments (see Chapter 6) are particularly striking in this respect. Only one-half of the haem a/Cu_A sites seem to be available for fast electron acceptance in the "resting" state of the enzyme. But the kinetic heterogeneity of cytochrome a in O_2-pulse and steady state experiments also point in the this direction.

In the following sections this idea is developed to some extent, and its main implications discussed.

A. Definition of the reciprocating site mechanism

Figure 7.2 shows a schematic representation of the proposed mechanism. We must emphasize that this figure should not be given too much structural significance. Large molecular movements are shown for illustrative purposes only.

It is proposed that the catalytic unit of cytochrome oxidase is an $(aa_3)_2$ dimer. During activity the dimer is asymmetric, so that one monomer is in the input and the companion monomer in the output state. Symmetrical states might exist under special conditions (see below). The input and output states correspond to the analogous, but more abstract counterparts in Fig. 7.1. Thus, when the monomer is in the input state (corresponding to a downward position in Fig. 7.2), electron transfer is possible only between cytochromes c and a, but not between a and the a_3/Cu_B centre. In this state the redox-linked acid/base group(s) is uniquely in protonic contact with the M phase.

Conversely, in the output state (upward position, Fig. 7.2) electron transfer is possible only between haem a and the a_3/Cu_B centre, but not between cytochromes c and a. The acid/base group(s) is (are) in contact with the C phase.

The overall catalytic mechanism is hence divided into two parts occurring simultaneously on the monomers in the input and output states. The two states are displaced by $180°$ with respect to the catalytic mechanism. After completion of the respective reaction steps there is a reciprocal conformational change by which both monomers switch states, the input monomer changing to the output state and vice versa, followed again by the catalytic steps specific for each state (Fig. 7.2). The conformational switch corresponds to the change in orientation discussed above for the more abstract model in Fig. 7.1. Figure 7.2 differs from this only by the contention that in the dimeric catalytic unit the changes in orientation occur simultaneously in the forward and backward directions. Due to its origins in Fig. 7.1, it is clear that the model of Fig. 7.2 also explains redox-linked proton translocation.

B. Evidence for and against dimeric functioning of cytochrome oxidase

The monomeric aa_3 unit has traditionally been considered to be the catalytic entity of cytochrome oxidase (see, for example, Lemberg, 1969; Azzi,

1980; Chapter 3; note that the aa_3 unit is sometimes referred to as the dimer in the earlier literature). More recently the enzyme has been found to be dimeric both in lateral membrane crystals and in solutions containing non-ionic detergents (Chapter 3). Henderson *et al.* (1977) suggested that the enzyme may also be dimeric *in situ* on the basis of such data. To our knowledge there is no unequivocal evidence of a homogeneous monomeric oxidase preparation with full catalytic potency. Monomeric preparations may be found to catalyse electron transfer (see Wilson *et al.*, 1980), which may still be consistent with our model. However, if they were shown to catalyse proton translocation, our model would be readily disproved.

Bisson and his collaborators (Bisson and Capaldi, 1980; Bisson *et al.*, 1980) recently reported that covalent cross-linking of a cytochrome *c* derivative to the high affinity site on the enzyme blocks electron transfer activity almost completely after binding of one cytochrome *c* per two aa_3 units. This "half-of-the-sites" effect provides strong evidence for functioning of the enzyme as a dimer (cf. also the kinetic "half-of-the-sites" effects described in Chapter 6). In fact, this kind of phenomenon has been considered characteristic not only of dimeric function, but more specifically, of so-called "flip-flop" mechanisms (see Lazdunski, 1972), of which the model in Fig. 7.2 is an example. "Half-of-the-sites" reactivity may be considered a special case of the phenomenon of negative co-operativity (see Levitzki and Koshland, 1976). Thus glycer-aldehyde-3-phosphate dehydrogenase, for example, shows antico-operative binding of NAD to the four subunits, and "half-of-the-sites" reactivity towards binding of other active site-directed agents. These two phenomena often occur simultaneously in the same protein. It is therefore of considerable interest to note that cytochrome oxidase is characterized by a high degree of antico-operativity with respect to the electron affinity of the haems (see Chapter 5). These phenomena have been interpreted as pure intramonomeric haem *a*/haem a_3 interactions. However, the occurrence of "half-of-the-sites" reactivity in the enzyme motivates a reinvestigation to find out whether some of the redox interactions may be of intermonomeric origin.

Wrigglesworth *et al.* (1973) found by titrating submitochondrial particles with cyanide that no inhibition took place until the cyanide/aa_3 ratio exceeded unity. This important finding provides evidence against co-operativity between a_3/Cu_B centres in the reaction with O_2. In the present context it would mean that the binding of cyanide to one monomer does not prevent the reciprocating transitions of the corresponding dimer. This finding is in a sense opposite to that made on covalent linkage of cytochrome *c* (see above), suggesting perhaps that binding of cytochrome *c* may be central in the reciprocating mechanism.

Brunori *et al.* (1979) showed that the "O_2-pulsed" oxidase (see also

Chapter 4 and 6) is formed from the reduced enzyme by combination of 1 mol of O_2 per aa_3 unit. This was suggested to indicate that the dimer has no particular role in enzyme activity. However, this finding is not inconsistent with the present model, but indicates only that there is no co-operation between two haem a_3/Cu_B centres in generation of the "pulsed" state (cf. the cyanide data above).

Note here, incidentally, that the finding that the redox state of cytochrome *a in toto* is a function of pH specifically in the M phase in cyanide-blocked mitochondria (Section II.C) is not in disagreement with the model. This may be surprising at first sight because the input and output states should exist simultaneously in the reciprocating site model. However, in the presence of cyanide the output state of cytochrome *a* cannot donate electrons to the a_3/Cu_B centre, which prevents observation of the dependence of pH in the C phase under such conditions.

We conclude that there are several structural and functional indications of a dimeric function of cytochrome oxidase. However, there are also data suggesting that the haem a_3/Cu_B centres catalyse O_2 reduction independently. This property is expected, actually, from the reciprocating site model in which no individual step in the overall reaction (apart from the "reorientation" steps) is catalysed by simultaneous co-operation of both monomers. For this reason it is also not inconceivable that a monomeric preparation can be isolated under suitable conditions that catalyses electron transfer and O_2 consumption. But according to the present model the monomeric enzyme could not catalyse H^+ translocation.

C. The nature of the monomer/monomer interactions

Intermonomeric interactions are essential in the proposed model. These have a reciprocal nature which means that when one specific catalytic event is turned on in one monomer, it is simultaneously turned off in the companion, and vice versa. Functions subject to this control are (i) the electronic contacts of haem *a* with cytochrome *c* and the a_3/Cu_B centre, respectively, and (ii) the protonic contacts of redox-linked acid/base groups with the aqueous C and M phases, respectively.

It may be of interest to consider how such reciprocal interactions may arise in the dimer of cytochrome oxidase, on the basis of the present structural knowledge.

In the membranous dimer the aa_3 monomers (or protomers) must be arranged head to head to fit the data on subunit topography, cytochrome *c* reactivity, etc. (Chapter 3). This is consistent with the apparent twofold rotational axis of symmetry along the long axis of the monomers, directed perpendicular to the plane of the membrane in so-called membrane crystals (Henderson *et al.*, 1977). From this it follows that the interactions between monomers are expected to be isologous along their surface of contact, which is directed perpendicularly through the membrane (see Monod *et al.*, 1965). In such a structure there would be point group symmetry along the contact surface, and the two monomers would be identical. In contrast to this, our model specifically requires a difference (asymmetry) between the monomers, i.e. the difference between input and output states (Fig. 7.2).

The important problem then arises as to the source of this asymmetry. The first possibility is that it is always present, i.e. it already exists in the fully oxidized, unperturbed enzyme. This may not be inconsistent with the electron microscopy and image reconstruction data if it is, for instance, achieved by a slight dislocation of the monomers with respect to one another along the symmetry axis, destroying exact point group symmetry. Such a case has been found, for example, in yeast hexokinase (Steitz *et al.*, 1976). On the other hand, the EPR properties of haem *a* are very homogeneous in the fully oxidized enzyme (Aasa *et al.*, 1976; Chapter 4), suggesting that the monomers may be symmetric in this "resting" state. Whether "oxygenated" and/or "pulsed" states differ from this by being asymmetric is an interesting possibility which, so far, lacks experimental support (see Section IV.D). However, the transient "$g = 5$" species observed by Beinert and his collaborators (see Chapters 4 and 6) is characterized by a split in the EPR resonances due to ferric haem *a* into two components. This might be due to the presence of two monomers associated asymmetrically.

A second possibility is that the necessary asymmetry is an induced property. The most obvious candidates causing asymmetry are binding of cytochrome *c* and reduction of the redox centres of the enzyme, respectively.

The two binding sites for cytochrome *c* (Chapters 3 and 6) are apparently present in one copy each per aa_3 unit (Mochan and Nicholls, 1972; Ferguson-Miller *et al.*, 1976, 1978*b*). Moreover, cross-linking of cytochrome *c* to the high affinity site produces an 1 : 1 complex between subunit II and the cytochrome (Briggs and Capaldi, 1978; Bisson *et al.*, 1978, 1980). Thus there is one high affinity site on each monomer without direct evidence of co-operation between these sites. The possibility that two *c* molecules bind with high affinity to one monomer inducing binding of two *c* molecules with low affinity to the companion monomer can be excluded. Hence, there is no evidence for the idea that the binding of cytochrome *c* would cause asymmetry between the monomers. Leigh *et al.* (1974) showed, in fact, that cytochrome *c* has no effect on the haems' thermodynamic parameters (see also Schroedl and Hartzell, 1977c), including the effects interpreted as strong antico-operative interactions (see Chapter 5 and below). Earlier findings to the contrary were due to the fact that redox equilibration among the oxidase's redox centres is extremely slow in the absence of cytochrome c.

Cytochrome c might thus have an accelerating (catalytic) effect on the reciprocating transitions. Control of catalytic activity by cytochrome c has also previously been considered on the basis of kinetic data (for recent initiated discussions, see Errede and Kamen, 1979; Ferguson-Miller et al., 1979). The recent data of Bisson et al. (1980) are interesting in this regard. Specific covalent binding of an arylazido cytochrome c derivative to the high affinity site prevented electron transfer even though the low affinity site remained free to bind non-derivatized cytochrome c. Moreover, as mentioned above, almost complete inhibition of electron transfer took place after covalent binding of only one c per dimer. If the low affinity site is indeed involved in electron input to the enzyme, as suggested by Ferguson-Miller et al. (1976, 1978b; cf. Krab and Slater, 1979), this result suggests interactions between low affinity and high affinity sites both within and between monomers.

As discussed in Chapter 5, the thermodynamic redox parameters of the haems are characterized by strong antico-operative interactions. These may be described as homotropic interactions since acceptance of electrons by the enzyme strongly decreases further electron affinity. Such effects may be unique for enzymes functioning effectively as dimers (Section IV.B; Lazdunski, 1972; Levitzki and Koshland, 1976). Such enzymes often show negative co-operativity in ligand binding both between protomers and between dimers. In cytochrome oxidase only intramonomeric a/a_3 interactions have been considered so far to explain the observed anticooperativity (Chapter 5). However, these effects might actually be due largely to intermonomeric a/a and, perhaps, a_3/a_3 interactions. Further work will be required to test this possibility rigorously. In the meantime we may consider the idea that the two cytochromes a may be identical (symmetric) in the "resting" state. Reduction of either one, which introduces asymmetry, may then have the effect of preventing reduction of the companion haem a by an intermonomeric reciprocal interaction. This would, incidentally, be an historically interesting analogy to the now abandoned "unitarian" hypothesis (see, for example, Sekuzu et al., 1959; Lemberg, 1969; Tiesjema et al., 1973), according to which the a/a_3 difference between two initially identical haems A in the monomer could be induced by reduction of either one.

It is, no doubt, a point in favour of the reciprocating site hypothesis that it may be able to provide a rational explanation of the strong haem/haem interactions in the enzyme, which up to now have remained completely enigmatic.

D. "Resting" and "pulsed" states of the dimer

Although explanation of phenomena that are not uniquely expected from the reciprocating site hypothesis means building of second-order hypotheses, some brief notes may be worthwhile concerning catalytically inactive and active states and the role of Cu_A (Section IV.E).

According to the present model, the anomalous slowness of electron transfer beyond cytochrome a in the "resting" state (Chapter 6) would be specifically due to slowness of the reciprocating input/output transition. As judged from the rate by which the slow (input) mode of cytochrome a is oxidized after an O_2 pulse to the reduced enzyme at room temperature ($t_{1/2}$ about 1 ms; see Chapter 6), this transition is indeed very much faster when initiated from the fully reduced state of the enzyme. It is in this latter condition that the "pulsed" enzyme is formed in which the kinetic restriction on electron transfer is largely relieved (see Chapter 6). It seems that the "resting" state is an anomalous conformation of the fully oxidized enzyme which is reached from an "active" oxidized state ("pulsed" or its equivalent) only after cessation of catalytic activity due to lack of electron donors. In contrast, cessation of activity in the reduced state, due to lack of O_2, apparently does not lead to an inactive state. It follows from these general considerations that any state of the enzyme other than that in which the haem a_3/Cu_B centre is stabilized in the Fe^{III}/Cu^{II} state, or its equivalent, would be "pulsed" in the sense that it will show high catalytic activity. This is consistent with the findings (Chapter 6) that partial reduction of the enzyme prior to the pulse of ferrocytochrome c leads to rapid kinetics. The potency of the fully reduced enzyme to catalyse rapid reciprocating movements, in contrast to the "resting" state, is also corroborated by the finding that added ferricytochrome c rapidly accepts *two* electrons per aa_3 unit (Chapter 6). In the present model this could not occur without the reciprocating movement of the monomers (see Fig. 7.2).

The molecular basis for the postulated block of reciprocating movements in the "resting" state is, of course, unknown, and any speculation to this end would immediately lead us to third-order hypotheses. Yet, one possibility is too evident from the above discussion to remain unmentioned, viz. that of symmetry. The "resting" state could thus be a unique state of the enzyme in which there is exact point group symmetry between the monomers. The transition to the catalytically active asymmetric states may be a slow reaction. The fully reduced enzyme may be asymmetric, which may be the reason behind the split absorption maximum of ferrous cytochrome a in the Soret band (Chapter 4). Alternatively, the reaction of a symmetric reduced state with O_2 may more easily lead to asymmetry and fast turnover than the transition from the "resting" state.

E. On the role of Cu_A in cytochrome oxidase

Cu_A is the most enigmatic redox centre of the enzyme for which no special function apart from that of a simple carrier of electrons has been obvious. The kinetics of enzyme oxidation with O_2 and reduction with ferrocyto-chrome c (Chapter 6) suggest that Cu_A may not be an obligatory link in the electron transfer sequence. Cu_A appears to equilibrate electronically with haem a both in the input and the output configurations of the latter. From these findings it appears as though Cu_A may function as a redox buffer of haem a, but the significance of such a function is not obvious.

It should be pointed out that, in the present models of Figs. 7.1 and 7.2, fully coupled H^+ translocation depends on the proviso that electrons are transferred obligatorily via the cyclic two-state function of cytochrome a. If, for instance, Cu_A were reduced by haem a in the input state (with release of the H^+ back into the M phase), and this were followed by the reciprocating input/output transition with Cu_A in the reduced state, oxidation of Cu_A by O_2 via haem a and the a_3/Cu_B centre in the output state may result. Such an electron transfer sequence would not result in proton translocation.

Under most conditions this situation may be unlikely, however, both for thermodynamic and kinetic reasons. Since the E_m of Cu_A is between 225 and 250 mV (Erecińska et al., 1971; Tsudzuki and Wilson, 1971), whereas the "functional" E_m of haem a is of the order of 275–300 mV or higher at pH 7 (Andréasson, 1975; Wikström et al., 1976; Brunori et al., 1979), reduction of the former by the latter may be thermodynamically unfavourable. It may also be approximated that, except for the resting state (Section VI.A, Chapter 6), the reciprocating transition is much faster ($t_{1/2} \sim 1$ ms at room temperature, see above) than is electron transfer from haem a to Cu_A in the input state (see Wilson et al., 1975). Hence, the input/output transition should usually occur before there is any significant reduction of Cu_A.

This situation may change significantly, however, in "high energy" or State 4 conditions where turnover is slow and there is an apparent decrease in the E_m of cytochrome a (Wikström et al., 1976). In such conditions it is possible that part of the electron flow through the enzyme may no longer be coupled to proton translocation by the proton pump. However, the energy-conserving function of the a_3/Cu_B centre still remains (Section II.E), so that the overall effect would be one of cutting the efficiency of energy conservation by 50%. A similar situation could then also arise in measurements of transients, e.g. fast oxidation of the fully reduced enzyme by O_2. Also here, part of the electron transfer in the first turnover may be "uncoupled" from proton translocation due to the pre-reduced state of Cu_A prior to the pulse.

The above considerations do not explain the function of Cu_A, unless a control of the efficiency of energy conservation by turning the pump on and off is important. Theoretical considerations suggest that such flexibility could be of great energetic significance in order to adjust mitochondrial energy conservation to an optimal efficiency (see Stucki, 1978).

The above speculations would suggest that Cu_A may be in a highly oxidized state under most aerobic conditions. This agrees with the findings of Beinert and Palmer (1965) in submitochondrial particles.

F. Structural implications and general significance

It was concluded (Section IV.C) that the model requires that exact point group symmetry is broken between the monomers. We favour the possibility that this is achieved by the presence of reducing equivalents in the enzyme's redox centres. Thus the asymmetric state may be the rule rather than the exception during catalytic activity. Also the fully oxidized state may exist in an asymmetric structure that may eventually relax to a symmetric ("resting") state if no electron donors are available.

One can visualize two equivalent states of the dimer in which exact point group symmetry is broken in precisely the same way, but with opposite roles of the two monomers (cf. Fig. 7.2). Catalysis involves oscillations between these two states, as shown schematically in Fig. 7.2. However, the nature of the symmetry distortion need not be a sliding movement parallel to the axis of symmetry, as depicted in the figure. Moreover, the actual movement is likely to be quite small with respect to the dimensions of the cytochrome oxidase molecule.

Another kind of symmetry distortion that seems more likely is rotation of the monomers in opposite directions around the symmetry axis. Note here, incidentally, that pivoting of monomers in opposite directions around an axis in the membrane plane, as often seen for pump models, cannot explain the reciprocal interactions in cytochrome oxidase. In such a case the symmetry would not be broken due to the "head-to-head" arrangement of the monomers, which would retain their equivalence throughout the movement. We should stress that molecular "movements" breaking the symmetry need not, of course, involve the entire monomers, as in Fig. 7.2, but may be much more local events.

We may finally extend our scope to the possibility that a reciprocating site type of mechanism may have a more general significance than for cytochrome oxidase alone. In this respect it is interesting that an analogous mechanism has been proposed for the H^+-translocating ATPase of mitochondria and chloroplasts, although based on a completely different set of experimental evidence (Kayalar et al., 1977; for a review see Wikström et al., 1981). The same is true for the Na^+/K^+-translocating ATPase, again quite independently (see Robinson and Flashner, 1979). These enzymes catalyse reactions that are analogous to that catalysed by cytochrome oxidase in that transport of ions across membranes is driven by a metabolic exergonic reaction. However, Lazdunski (1972) has listed more than 15 different non-transport enzymes that are likely to catalyse a "flip-flop" type of mechanism analogous to that proposed here for cytochrome oxidase. These included malate dehydrogenase, for which oxidoreduction and substrate binding was suggested to occur alternately on different subunits, the prototype of substrate level phosphorylation, viz. glyceraldehyde phosphate dehydrogenase, and methionyl-tRNA synthetase. Interestingly, in 1968 Harada and Wolfe had already proposed a "reciprocating compulsory order mechanism" for malate dehydrogenase purely on the basis of kinetic anomalies observed (cf. Chapter 6).

It is clear from this that if cytochrome oxidase indeed functions by a reciprocating site type of mechanism, it may by no means be unique in its principle of catalysis. What may be novel, however, is the idea that a reciprocating site mechanism may have special significance in enzymes that catalyse biological energy transduction (Wikström et al., 1981). Lazdunski (1972) considered four kinds of special advantage of so-called flip-flop mechanisms, viz. (i) thermodynamic coupling between successive steps in the catalytic mechanism; (ii) a special evolutionary advantage; (iii) an advantage in the function of enzyme assemblies; (iv) an advantage in inhibitory effects of substrates and effectors.

For energy-transducers such as cytochrome oxidase the advantage in (i) may be crucial, as a thorough kinetic and thermodynamic analysis of Figs 7.1 and 7.2 indicates. The intermonomeric interaction may also assure tightness of coupling with little chance of molecular "slipping" (though some modulation of this property might occur). In glyceraldehyde phosphate dehydrogenase, one of the candidates for a flip-flop mechanism, this is achieved so that oxidation to phosphoglycerate cannot take place without phosphorylation of the product. It is intriguing that if this analogy turns out to be correct, there would, after all, exist a clear similarity between mechanisms of oxidative and substrate-level phosphorylation, although in a sense not previously anticipated.

Also, the advantage in (iii) might find a special meaning. Thus transport enzymes could be looked upon as "tightly associated" enzyme assemblies, consisting of separate catalytic units for the chemical and the physical (transport) reaction.

V. Conclusions

Cytochrome oxidase conserves redox energy by redox-linked proton translocation and by reducing dioxygen to water with electrons derived from one side and protons from the other side of the mitochondrial membrane. Thermodynamically, the overall energy transduction is equivalent to translocation of $2H^+/e^-$ across the membrane.

Cytochrome a is specifically involved in the proton pump mechanism, in which $1H^+/e^-$ is translocated. The haem a_3/Cu_B accepts the electrons derived from cytochrome c and protons deriving from the M phase in the reduction of dioxygen to water. This results in uptake of another H^+/e^- from the M side and the translocation of another electrical charge across the membrane.

Subunit III may be specifically involved in the proton pump mechanism by functioning as a gated H^+ channel. There are some, albeit weak, indications that residues on, or in the vicinity of, haem a may be involved in the proton pump mechanism.

A tentative mechanism, the reciprocating site hypothesis, was presented in which cytochrome oxidase functions as a dimer. General implications of this model were discussed.

Epilogue: Why a proton pump?

One may finally be motivated to ask to what extent conclusions made with respect to cytochrome oxidase structure and function are strictly limited to this enzyme, and to what extent they might be more generally applicable to biological electron transfer and energy transduction.

The questions that have concerned us most strongly in this vein are why cytochrome oxidase functions as a proton pump and further, whether this principle may also be applicable to other systems. The alternative mechanisms of redox loop and electron translocator suggested by Robertson (1960) and by Mitchell (1961, 1979) seem to be well established for energy transduction by nitrate reductase and some other bacterial systems (Jones et al., 1980; Kröger, 1978). Also, generation of electrochemical proton gradients by photosystems I and II, and by bacteriochlorophyll, are almost certainly due to so-called direct coupling. Yet other segments of photosynthetic electron transfer, such as the cytochrome $b_6 f$ and bc_2 systems of chloroplasts and photosynthetic bacteria, might function as proton pumps, although this has by no means been demonstrated unequivocally (see Wikström et al., 1981).

In mitochondria the structural, and to some extent functional, analogies between the bc_1 and aa_3 complexes might indicate that the former also functions as a proton pump. However, at the time of writing it has not been possible to distinguish experimentally between a redox loop mechanism and a proton pump in this complex. The NADH dehydrogenase complex is even more enigmatic in this respect. The H^+/e^- ratio of more than unity for proton translocation (Pozzan et al., 1979) supports a proton pump type of mechanism, but this finding is still subject to some controversy (Mitchell and Moyle, 1979; Wikström and Krab, 1980). Although this system has been much less studied recently than the so-called cytochrome chain, it may be of particular interest from an evolutionary point of view. If the iron–sulphur proteins that are abundant in this complex, and may be involved in energy transduction (Ohnishi, 1975), may be considered as "primordial cytochromes" (Baltscheffsky, 1974, 1978), a proton pump mechanism analogous to that in cytochrome oxidase may more easily be envisaged.

The energy-dependent nicotinamide dinucleotide transhydrogenase, which is auxiliary to the respiratory chain, almost certainly functions as a

171

redox-linked proton pump (Rydström, 1977, 1979). Here it may be difficult *a priori* to draw any close parallel to cytochrome oxidase, but one possibility is discussed below.

Bacteriorhodopsin constitutes a proton pump that is unique in that it is driven directly by light energy (and not through oxidoreductions as in photosynthesis). It might yet turn out that there are analogies between this system and that of cytochrome oxidase, although the evidence in this direction is still scanty (Chapter 7).

All in all it seems to us that Nature may utilize different principles for conservation of available energy. The extent to which these systems are analogous to one another, if at all, cannot be evaluated with confidence at present.

But then the next question emerges, which is closely relevant to the topic of this book. No doubt pure electron translocation, which may be easily simulated in a simple artificial system (Hinkle, 1973), and a redox loop are both chemically and structurally simpler mechanisms of generation of $\Delta\bar{\mu}_{H^+}$ than is a proton pump. Why then has the proton pump mechanism evolved? Here the reader should be reminded that an ion pump such as the proton pump of cytochrome oxidase is, of course, not an isolated principle in biology but is widely utilized in various transport systems. The ones that are functionally closest to cytochrome oxidase are perhaps the H^+-translocating ATPases of the mitochondrial, bacterial and photosynthetic membranes, but other ion pumps also function along general principles of indirect coupling (e.g. the Na^+/K^+-ATPase, Ca^{2+}-ATPases, etc.). The question is then rather why this principle has been utilized in linkage with some redox reactions such as that of cytochrome oxidase and, perhaps, transhydrogenase, although these have the unique potential of forming directly coupled redox loops or electron translocating systems of much greater simplicity?

It is perhaps surprising that at least for cytochrome oxidase it seems to be easy to answer this question. The answer may be twofold. First, due to the large redox span covered, mere electron translocation would conserve only a small fraction of the available energy (see Chapter 2). Thus a hydrogen-translocating limb of the redox loop would also be required to make it possible for electrons to traverse the membrane dielectric twice. However, there is no biological hydrogen carrier with a high enough E_m to fit this function at the high operative redox potentials of the cytochrome oxidase reaction. We suggest that it may be for these reasons that a proton pump mechanism has evolved.

In view of the striking similarities between structure and, possibly, function of the F_0 segment of H^+-ATPase on the one hand and subunit III of cytochrome oxidase on the other, it may not be inconceivable that the proton pump of the latter has evolved following the pattern of the ATPase. If the reciprocating site mechanism proposed for cytochrome oxidase turns out to be essentially correct, there may be another interesting analogy to the H^+-ATPases, which have been suggested to function similarly on the basis of very different and independent evidence (Kayalar *et al.*, 1977). It is possible that this opens an analogy to the energy-linked transhydrogenase (see above), which may well be dimeric *in situ* and shows several allosteric properties (Rydström, 1977; Höjeberg and Rydström, 1977). In view of the potential ability of reciprocal or antico-operative interactions between similar catalytic units to provide thermodynamic coupling between elementary steps of the overall enzymic process (see Lazdunski, 1972; Levitzki and Koshland, 1976), and the increasing number of enzymes in which such interactions seem to be discovered, it seems to us that this principle is a very basic one in enzyme reactions. It is therefore not too far-fetched to consider the possibility that this principle may be utilized in the very reactions that we usually classify as biological energy transductions (Wikström *et al.*, 1981), including the coupling between chemical and physical reactions, as is the case in active transport for example. Whether this is so can only be assessed by future experimentation.

References

Aasa, R., Albracht, S. P. J., Falk, K.-E., Lanne, B. and Vänngård, T. (1976) *Biochim. Biophys. Acta* **422**, 260–272

Alben, R., Altschuld, F., Fiamingo, F. and Moh, P. (1981) in "Interaction Between Iron and Proteins in Oxygen and Electron Transport" (C. Ho and W. C. Eaton, eds), Elsevier, New York, in press

Albracht, S. P. J., Van Verseveld, H. W., Hagen, W. R. and Kalkman, M. L. (1980) *Biochim. Biophys. Acta* **593**, 173–186

Alexandre, A. and Lehninger, A. L. (1979) *J. Biol. Chem.* **254**, 11555–11560

Anderson, J. L., Kuwana, T. and Hartzell, C. R. (1976) *Biochemistry* **15**, 3847–3855

Anderson, S., Bankier, A. T., Barrell, B. G., de Bruijn, M. H. L., Coulson, A. R., Drouin, J., Eperon, I. C., Nierlich, D. P., Roe, B. A., Sanger, F., Schreier, P. H., Smith, A. J. H., Staden, R. and Young, I. G. (1981a) *Nature* **290**, 457–465

Anderson, S., de Bruijn, M. H. L., Coulson, A. R., Eperon, I. C., Sanger, F. and Young, I. G. (1981b) In preparation

Andréasson, L.-E. (1975) *Eur. J. Biochem.* **53**, 591–597

Andréasson, L.-E., Malmström, B. G., Strömberg, C. and Vänngård, T. (1972) *FEBS Lett.* **28**, 297–301

Antonini, E., Brunori, M., Greenwood, C., Malmström, B. G. and Rotilio, G. C. (1971) *Eur. J. Biochem.* **23**, 396–400

Antonini, E., Brunori, M., Greenwood, C., Colosimo, A. and Wilson, M. T. (1977) *Proc. Natl Acad. Sci. U.S.A.* **74**, 3128–3132

Ariano, B. H. and Azzi, A. (1980a) *Biochem. Biophys. Res. Commun.* **93**, 478–485

Ariano, B. H. and Azzi, A. (1980b) in "1st European Bioenergetics Conference: Short Reports", Patron Editore, Bologna, pp. 93–94

Artzatbanov, V. Y., Konstantinov, A. A. and Skulachev, V. P. (1978) *FEBS Lett.* **87**, 180–185

Azzi, A. (1980) *Biochim. Biophys. Acta* **594**, 231–252

Azzi, A. and Casey, R. P. (1979) *Mol. Cell Biochem.* **28**, 169–184

Azzone, G. F., Pozzan, T. and Di Virgilio, F. (1979) *J. Biol. Chem.* **254**, 10206–10212

Babcock, G. T. and Chang, C. K. (1979) *FEBS Lett.* **97**, 358–362

Babcock, G. T. and Salmeen, I. (1979) *Biochemistry* **18**, 2493–2498

Babcock, G. T., Vickery, L. E. and Palmer, G. (1976) *J. Biol. Chem.* **251**, 7907–7919

Babcock, G. T., Vickery, L. E. and Palmer, G. (1978) *J. Biol. Chem.* **253**, 2400–2411

Babcock, G. T., Van Steelan, J., Palmer, G., Vickery, L. E. and Salmeen, I. (1979) in "Cytochrome Oxidase" (T. E. King, Y. Orii, B. Chance and K. Okunuki, eds), Elsevier, Amsterdam, pp. 105–115

Babcock, G. T., Callahan, P. M., Ondrias, M. R. and Salmeen, I. (1981) *Biochemistry*, **20**, 959–966

Baltscheffsky, H. (1974) in "Dynamics of Energy-transducing Membranes" (L. Ernster, R. W. Estabrook and E. C. Slater, eds), Elsevier, Amsterdam, pp. 21–27

Baltscheffsky, H. (1978) in "Energy Conservation in Biological Membrane" (G. Schäfer and M. Klingenberg, eds), Springer Verlag, Berlin, pp. 3–18

Barlow, C. and Erecińska, M. (1979) *FEBS Lett.* **98**, 9–12.

Barlow, C. H., Maxwell, J. C., Wallace, W. J. and Caughey, W. S. (1973) *Biochem. Biophys. Res. Commun.* **55**, 91–95

Barrell, B. G., Bankier, A. T. and Drouin, J. (1979) *Nature* **282**, 189–194

Beechey, R. B., Roberton, A. M., Holloway, C. T. and Knight, I. G. (1967) *Biochemistry* **6**, 3867–3879

Beinert, H. and Palmer, G. (1965) in "Oxidases and Related Redox Systems" (T. E. King, H. S. Mason and M. Morrison, eds), Wiley, New York, pp. 567–585

Beinert, H. and Shaw, R. W. (1977) *Biochim. Biophys. Acta* **504**, 187–199

Beinert, H., Hansen, R. E. and Hartzell, C. R. (1976) *Biochim. Biophys. Acta* **423**, 339–355

Beinert, H., Shaw, R. W. and Hansen, R. E. (1979) in "Cytochrome Oxidase" (T. E. King, Y. Orii, B. Chance and K. Okunuki, eds), Elsevier/North-Holland, Amsterdam, pp. 139–152

Beinert, H., Shaw, R. W., Hansen, R. E. and Hartzell, C. R. (1980) *Biochim. Biophys. Acta* **591**, 458–470.

Bernstein, J. D., Bucher, J. R. and Penniall, R. (1978) *J. Bioenerg. Biomembr.* **10**, 59–74

Bertrand, H. and Werner, S. (1979) *Eur. J. Biochem.* **98**, 9–18

Bienfait, H. F. (1975) Reversibility of site 3 phosphorylation. Ph.D. thesis, University of Amsterdam

Bill, K., Casey, R. P., Broger, C. and Azzi, A. (1980) *FEBS Lett.* **120**, 248–250

Birchmeier, W., Kohler, C. E. and Schatz, G. (1976) *Proc. Natl. Acad. Sci. U.S.A.* **73**, 4334–4338

Bisson, R. and Capaldi, R. A. (1980) in "1st European Bioenergetics Conference: Short Reports", Patron Editore, Bologna pp. 103–104

Bisson, R., Gutweniger, H., Montecucco, C., Colonna, R., Zanotti, A. and Azzi, A. (1977) *FEBS Lett.* **81**, 147–150

Bisson, R., Azzi, A., Gutweniger, H., Colonna, R., Montecucco, C. and Zanotti, A. (1978) *J. Biol. Chem.* **253**, 1874–1880

Bisson, R., Montecucco, C., Gutweniger, H. and Azzi, A. (1979) *J. Biol. Chem.* **254**, 9962–9965

Bisson, R., Jacobs, B. and Capaldi, R. (1980) *Biochemistry* **19**, 4173–4178

Blasie, J. K., Erecińska, M., Samuels, S. and Leigh, J. S. Jr (1978) *Biochim. Biophys. Acta* **501**, 33–52

Blokzijl-Homan, M. F. J. and Van Gelder, B. F. (1971) *Biochim. Biophys. Acta* **234**, 493–498

Blum, H., Harmon, H. J., Leigh, J. S., Salerno, J. C. and Chance, B. (1978) *Biochim. Biophys. Acta* **502**, 1–10

Blumberg, W. E. and Peisach, J. (1979) in "Cytochrome Oxidase" (T. E. King, Y. Orii, B. Chance and K. Okunuki, eds), Elsevier/North-Holland, Amsterdam, pp. 153–159

Boelens, R. and Wever, R. (1980) *FEBS Lett.* **116**, 223–226

Bonitz, S. G., Coruzzi, G., Thalenfeld, B. E., Tzagoloff, A. and Macino, G. (1980) *J. Biol. Chem.* **255**, 11927–11941

Boonman, J. (1979) Subunits of beef heart cytochrome oxidase. Ph.D. dissertation, Laboratory of Biochemistry, B. C. P. Jansen Institute, University of Amsterdam.

Borst, P. and Grivell, L. A. (1978) *Cell* **15**, 705–723

Boyer, P. D. (1975) *FEBS Lett.* **58**, 1–6

Boyer, P. D. (1980) in "1st European Bioenergetics Conference: Short Reports", Patron Editore, Bologna, pp. 133–134

Boyer, P. D., Chance, B., Ernster, L., Mitchell, P., Racker, E. and Slater, E. C. (1977) *Annu. Rev. Biochem.* **46**, 955–1026

Brautigan, B. L., Ferguson-Miller, S. and Margoliash, E. (1978) *J. Biol. Chem.* **253**, 130–139

Briggs, M. M. and Capaldi, R. A. (1977) *Biochemistry* **16**, 73–77

Briggs, M. M. and Capaldi, R. A. (1978) *Biochem. Biophys. Res. Commun.* **80**, 553–559

Briggs, M., Kamp, P. F., Robinson, N. C. and Capaldi, R. A. (1975) *Biochemistry* **14**, 5123–5128

Brill, A. S. and Williams, R. J. P. (1961) *Biochem. J.* **78**, 246–253

Brittain, T. and Greenwood, C. (1976) *Biochem. J.* **155**, 453–455

Brudwig, G. W., Stevens, T. H. and Chan. S. I. (1980) *Biochemistry* **19**, 5275–5285

Brunori, M., Colosimo, A., Rainoni, O., Wilson, M. T. and Antonini, E. (1979) *J. Biol. Chem.* **254**, 10769–10775

Bucher, J. R. and Penniall, R. (1975) *FEBS Lett.* **60**, 180–184

Buse, G. and Steffens, G. J. (1978) *Hoppe-Seyler's Z. Physiol. Chem.* **359**, 1005–1009

Buse, G., Steffens, G. J. and Steffens, G. C. M. (1978) *Hoppe-Seyler's Z. Physiol. Chem.* **359**, 1011–1013

Buse, G., Steffens, G. C. M., Steffens, G. J., Sacher, R. and Erdweg, M. (1980) in "1st European Bioenergetics Conference: Short Reports", Patron Editore, Bologna, pp. 41–42

Cabral, F. and Love, B. (1972) *Biochim. Biophys. Acta* **283**, 181–186

Callahan, P. M. and Babcock, G. T. (1980) *Biochemistry* **20**, 952–958

Capaldi, R. A. (1973) *Biochim. Biophys. Acta* **303**, 237–241

Capaldi, R. A. (1981) in "Interaction Between Iron and Proteins in Oxygen and Electron Transport" (C. Ho and W. C. Eaton, eds), Elsevier, New York, in press

Capaldi, R. A. and Briggs, M. (1976) in "The Enzymes of Biological Membranes" Vol. 4 (A. Martonosi, ed.), Wiley, New York, pp. 87–102.

Capaldi, R. A. and Hayashi, H. (1972) *FEBS Lett.* **26**, 261–263

Capaldi, R. A., Bell, R. L. and Branchek, T. (1977) *Biochem. Biophys. Res. Commun.* **74**, 425–433

Carithers, R. and Palmer, G. (1981) in "Interaction Between Iron and Proteins in Oxygen and Electron Transport (C. Ho and W. C. Eaton, eds), Elsevier, New York, in press

Carroll, R. C. and Racker, E. (1977) *J. Biol. Chem.* **252**, 6981–6990

Casey, R. P., Chappell, J. B. and Azzi, A. (1979) *Biochem. J.* **182**, 149–156

Casey, R. P., Thelen, M. and Azzi, A. (1980) *J. Biol. Chem.* **255**, 3994–4000

Cattell, K. J., Lindop, C. R., Knight, I. G. and Beechey, R. B. (1971) *Biochem. J.* **125**, 169–177

Caughey, W. S., Barlow, C. H., Maxwell, J. C., Volpe, J. A. and Wallace, W. J. (1975a) *Ann. N.Y. Acad. Sci.* **244**, 1–9

Caughey, W. S., Smythe, G. A., O'Keefe, D. H., Maskasky, J. E. and Smith, M. L. (1975b) *J. Biol. Chem.* **250**, 7602–7622

Caughey, W. S., Wallace, W. J., Volpe, J. A. and Yoshikawa, S. (1976) in "The Enzymes", 3rd edn, Vol. 13C (P. D. Boyer, ed.), Academic Press, New York, pp. 299–344

Caughey, W. S., Choc, M. G. and Houtchens, R. A. (1979) in "Biochemical and Clinical Aspects of Oxygen" (W. S. Caughey, ed.), Academic Press, New York, pp. 1–17

Cerletti, N. and Schatz, G. (1979) *J. Biol. Chem.* **254**, 7746–7751

Chan, S. H. P. and Tracy, R. P. (1978) *Eur. J. Biol.* **89**, 595–605

Chan, S. I., Bocian, D. F., Brudwig, G. W., Morse, R. H. and Stevens, T. H. (1978) in "Frontiers of Biological Energetics" (P. L. Dutton *et al.*, eds), Academic Press, New York, pp. 883–888

Chan, S. I., Bocian, D. F., Brudwig, G. W., Morse, R. H. and Stevens, T. H. (1979) in "Cytochrome Oxidase" (T. E. King, Y. Orii, B. Chance and K. Okunuki, eds), Elsevier/North-Holland, Amsterdam, pp. 177–188

Chance, B. (1952) *Arch. Biochem. Biophys.* **41**, 404–415

Chance, B. and Erecińska, M. (1971) *Arch. Biochem. Biophys.* **143**, 675–687

Chance, B. and Leigh, J. S. Jr (1977) *Proc. Natl Acad. Sci. U.S.A.* **74**, 4777–4780

Chance, B. and Schindler, F. (1965) in "Oxidases and Related Redox Systems" Vol. 2 (T. E. King *et al.*, eds), Wiley, New York, pp. 921–929

Chance, B. and Williams, G. R. (1955) *Advan. Enzymol.* **217**, 409–426

Chance, B., Saronio, C. and Leigh, J. S. (1975) *J. Biol. Chem.* **250**, 9226–9237

Chance, B., Leigh, J. S. Jr and Waring, A. (1977) in "Structure and Function of Energy-Transducing Membranes" (K. van Dam and B. F. van Gelder, eds), Elsevier/North-Holland, Amsterdam, pp. 1–10

Chance, B., Saronio, C., Waring, A. and Leigh, J. S. Jr (1978) *Biochim. Biophys. Acta* **503**, 37–55

Chance, B., Saronio, C. and Leigh, J. S. (1979) *Biochem. J.* **177**, 931–941

Chapman, D., Gomez-Fernandez, J. C. and Goni, F. M. (1979) *FEBS Lett.* **98**, 211–223

Chuang, T. F. and Crane, F. L. (1971) *Biochem. Biophys. Res. Commun.* **42**, 1076–1081

Churg, A. K., Glick, H. A. Zelano, J. A. and Makinen, M. W. (1979) in "Biochemical and Clinical Aspects of Oxygen" (W. S. Caughey, ed.), Academic Press, New York, pp. 125–139

Clark, W. M. (1960) "Oxidation-Reduction Potentials of Organic Systems", Williams and Wilkins, Baltimore, Md

Clore, G. M. and Chance, B. (1978a) *Biochem. J.* **173**, 799–810

Clore, G. M. and Chance, B. (1978b) *Biochem. J.* **173**, 811–820

Clore, G. M., Andréasson, L.-E., Karlsson, B., Aasa, R. and Malmström, B. G. (1980a) *Biochem. J.* **185**, 139–154

Clore, G. M., Andréasson, L.-E., Karlsson, B., Aasa, R. and Malmström, B. G. (1980b) *Biochem. J.* **185**, 155–167

Coin, J. T. and Hinkle, P. C. (1979) in "Membrane Bioenergetics" (C. P. Lee *et al.*, eds), Addison-Wesley, Reading, Mass., pp. 405–412

Colman, P. M., Freeman, H. C., Guss, J. M., Murata, M., Norris, V. A., Ramshaw, J. A. M. and Venkatappa, M. P. (1978) *Nature* **272**, 319–324

Coruzzi, C. and Tzagoloff, A. (1979) *J. Biol. Chem.* **254**, 9324–9330

Deatherage, J. F., Henderson, R. and Capaldi, R. A. (1980) in "Interaction

Between Iron and Proteins in Oxygen and Electron Transport" (C. Ho and W. C. Eaton, eds), Elsevier, New York, in press

Degn, H. and Wohlrab, H. (1971) *Biochim. Biophys. Acta* **245**, 347–355

Denis, M. (1977) *FEBS Lett.* **84**, 296–298

Denis, M. (1981) *Biochim. Biophys. Acta* **634**, 30–40

Denis, M., Neau, E., Agalidis, I. and Pajot, P. (1981) in "1st European Bioenergetics Conference: Short Reports", Patron Editore, Bologna, pp. 81–82

DePierre, J. W. and Ernster, L. (1977) *Annu. Rev. Biochem.* **46**, 201–262

DerVartanian, D. V., Lee, I. Y., Slater, E. C. and Van Gelder, B. F. (1974) *Biochim. Biophys. Acta* **347**, 321–327

DeVault, D. (1971) *Biochim. Biophys. Acta* **226**, 193–199

DeVault, D. and Chance, B. (1966) *Biophys. J.* **6**, 825–847

Dickerson, R. E. and Timkovich, R. (1975) in "The Enzymes", 3rd edn, Vol. 11 (P. D. Boyer, ed.), Academic Press, New York, pp. 397–547

Dockter, M. E., Steinemann, A. Schatz, G. (1978) *J. Biol. Chem.* **253**, 311–317

Downer, N. W., Robinson, N. C. and Capaldi, R. A. (1976) *Biochemistry* **15**, 2930–2936

Dutton, P. L. (1971) *Biochim. Biophys. Acta* **226**, 63–80

Dutton, P. L. and Wilson, D. F. (1974) *Biochim. Biophys. Acta* **346**, 165–212

Dutton, P. L., Wilson, D. F. and Lee, C. P. (1970) *Biochemistry* **9**, 5077–5082

Dutton, P. L., Leigh, J. S., Jr and Scarpa, A. (eds) (1978) "Frontiers of Biological Energetics", Vols 1 and 2, Academic Press, New York

Ebner, E., Mason, T. L. and Schatz, G. (1973) *J. Biol. Chem.* **248**, 5369–5378

Eglinton, D. G., Johnson, M. K., Thomson, A. J., Gooding, P. E. and Greenwood, C. (1980) *Biochem. J.* **191**, 319–531

Ehrenberg, A. and Yonetani, T. Y. (1961) *Acta Chem. Scand.* **15**, 1071–1080

Erecińska, M. (1975) *Arch. Biochem. Biophys.* **169**, 199–208

Erecińska, M. (1977) *Biochem. Biophys. Res. Commun.* **76**, 495–501

Erecińska, M. and Chance, B. (1972) *Arch. Biochem. Biophys.* **151**, 304–315

Erecińska, M. and Wilson, D. F. (1978) *Arch. Biochem. Biophys.* **188**, 1–14

Erecińska, M., Chance, B. and Wilson, D. F. (1971) *FEBS Lett.* **16**, 284–286

Erecińska, M., Wilson, D. F., Sato, N. and Nicholls, P. (1972) *Arch. Biochem. Biophys.* **151**, 188–193

Erecińska, M., Blasie, J. K. and Wilson, D. F. (1977) *FEBS Lett.* **76**, 235–239

Erecińska, M., Wilson, D. F. and Blasie, J. K. (1978a) *Biochim. Biophys. Acta* **501**, 53–62

Erecińska, M., Wilson, D. F. and Blasie, J. F. (1978b) *Biochim. Biophys. Acta* **501**, 63–71

Erecińska, M., Wilson, D. F. and Blasie, J. K. (1979) *Biochim. Biophys. Acta* **545**, 352–364

Errede, B. and Kamen, M. D. (1978) *Biochemistry* **17**, 1015–1027

Errede, B. and Kamen, M. D. (1979) in "Cytochrome Oxidase" (T. E. King, Y. Orii, B. Chance and K. Okunuki, eds), Elsevier/North-Holland, Amsterdam, pp. 269–280

Errede, B., Haight, G. P. and Kamen, M. D. (1976) *Proc Natl Acad. Sci. U.S.A.* **73**, 113–117

Eytan, G. D. and Broza, R. (1978) *J. Biol. Chem.* **253**, 3196–3202

Eytan, G. D. and Schatz, G. (1975) *J. Biol. Chem.* **250**, 767–774

Eytan, G. D., Carroll, R. C., Schatz, G. and Racker, E. (1975) *J. Biol. Chem.* **250**, 8598–8603

Falk, J. E., Lemberg, R. and Morton, R. K. (eds) (1961) "Haematin Enzymes", Pergamon Press, Oxford.

Falk, K.-E., Vänngård, T. and Ångström, J. (1977) *FEBS Lett.* **75**, 23–27

Fee, J. A., Choc, M. G., Findling, K. L., Lorence, R. and Yoshida, T. (1980) *Proc. Natl Acad. Sci. U.S.A.* **77**, 147–151

Ferguson-Miller, S., Brautigan, D. L. and Margoliash, E. (1976) *J. Biol. Chem.* **251**, 1104–1115

Ferguson-Miller, S., Brautigan, D. L. and Margoliash, E. (1978a) *J. Biol. Chem.* **253**, 140–148

Ferguson-Miller, S., Brautigan, D. L. and Margoliash, E. (1978b) *J. Biol. Chem.* **253**, 149–159

Ferguson-Miller, S., Weiss, H., Spock, S. H., Brautigan, D. L., Osheroff, N. and Margoliash, E. (1979) in "Cytochrome Oxidase" (T. E. King, Y. Orii, B. Chance and K. Okunuki, eds), Elsevier/North-Holland, Amsterdam, pp. 281–292

Florkin, M. (1975) in "Comprehensive Biochemistry", Vol. 31 (M. Florkin and E. H. Stotz, eds), Elsevier, Amsterdam, pp. 187–235

Fowler, L. R., Richardson, S. H. and Hatefi, Y. (1962) *Biochim. Biophys. Acta* **64**, 170–173

Fox, T. (1979) *Proc. Natl Acad. Sci. U.S.A.* **76**, 6534–6538

Freedman, J. A., Tracy, R. P. and Chan, S. H. P. (1979) *J. Biol. Chem.* **254**, 4305–4308

Frey, T. G., Chan, S. H. P. and Schatz, G. (1978) *J. Biol. Chem.* **253**, 4389–4395

Froncisz, W., Scholes, C. P., Hyde, J. S., Wei, Y.-H., King, T. E., Shaw, R. W. and Beinert, H. (1979) *J. Biol. Chem.* **254**, 7482–7484

Fry, M. (1979) *Biochem. Biophys. Res. Commun.* **90**, 1119–1124

Fry, M. and Green, D. E. (1980) *Biochem. Biophys. Res. Commun.* **95**, 1529–1535

Fry, M., Vande Zande, H. and Green, D. E. (1978) *Proc. Natl Acad. Sci. U.S.A.* **75**, 5908–5911

Fuller, S., Capaldi, R. A. and Henderson, R. (1979) *J. Mol. Biol.* **134**, 305–327

George, P. (1953) *J. Biol. Chem.* **201**, 413–426

George, P. (1965) in "Oxidases and Related Redox Systems", Vol. 1 (T. E. King *et al.*, eds), Wiley, New York, pp. 3–33

George, P. and Irvine, D. H. (1955) *Biochem. J.* **60**, 596–601

George, P., Beetlestone, J. and Griffith, J. S. (1961) in "Haematin Enzymes", part 1 (J. E. Falk *et al.*, eds), Pergamon Press, Oxford, pp. 105–139

Gibson, Q. H. and Greenwood, C. (1963) *Biochem. J.* **86**, 541–554

Gibson, Q. H. and Greenwood, C. (1965) *J. Biol. Chem.* **240**, 2694–2698

Gibson, Q. H., Greenwood, C., Wharton, D. C. and Palmer, G. (1965) *J. Biol. Chem.* **240**, 888–894

Gilmour, M. V., Wilson, D. F. and Lemberg, R. (1967) *Biochim. Biophys. Acta* **143**, 487–499

Gilmour, M. V., Lemberg, M. R. and Chance, B. (1969) *Biochim. Biophys. Acta* **172**, 37–51

Green, D. E. and Vande Zande, E. (1981) *Biochem. Biophys. Res. Commun.* **98**, 635–641

Greenaway, F. T., Chan, S. H. P. and Vincow, G. (1977) *Biochim. Biophys. Acta* **490**, 62–78

Greenwood, C. and Gibson, Q. H. (1967) *J. Biol. Chem.* **242**, 1782–1787

Greenwood, C., Wilson, M. T. and Brunori, M. (1974) *Biochem. J.* **137**, 205–215

Greenwood, C., Brittain, T., Wilson, M. and Brunori, M. (1976) *Biochem. J.* **157**, 591–598

Griffith, J. S. (1971) *Mol. Phys.* **21**, 141–143

Gurd, F. R. N., Falk, K.-E., Malmström, B. G. and Vänngård, T. (1967) *J. Biol. Chem.* **242**, 5724–5730

Gutteridge, S., Winter, D. B., Bruyninckx, W. J. and Mason, H. S. (1977) *Biochem. Biophys. Res. Commun.* **78**, 945–951

Hackenbrock, C. R. (1977) in "Structure of Biological Membranes" (S. Abrahamsson and I. Pascher, eds), Plenum Press, London, pp. 199–234

Hackenbrock, C. R. and Miller-Hammon, K. (1975) *J. Biol. Chem.* **250**, 9185–9197

Harada, K. and Wolfe, R. G. (1968) *J. Biol. Chem.* **243**, 4131–4137

Hare, J. F., Ching, E. and Attardi, G. (1980) *Biochemistry* **19**, 2023–2030

Hartzell, C. R. and Beinert, H. (1974) *Biochim. Biophys. Acta* **368**, 318–338

Hartzell, C. R. and Beinert, H. (1976) *Biochim. Biophys. Acta* **423**, 323–338

Hartzell, C. R., Hansen, R. E. and Beinert, H. (1973) *Proc. Natl Acad. Sci. U.S.A.* **70**, 2477–2481

Hartzell, C. R., Beinert, H. Van Gelder, B. F. and King, T. E. (1978) *Methods Enzymol.* **53**, 54–66

Helenius, A. and Simons, K. (1975) *Biochim. Biophys. Acta* **415**, 29–79

Henderson, R., Capaldi, R. A. and Leigh, J. S. (1977) *J. Mol. Biol.* **112**, 631–648

Hillman, K. and Wainio, W. W. (1977) *J. Bioenerg. Biomembr.* **9**, 181–193

Hinkle, P. (1973) *Fed. Proc.* **32**, 1988–1992

Hinkle, P. and Mitchell, P. (1970) *J. Bioenerg.* **1**, 45–60.

Hinkle, P. C., Kim, J. J. and Racker, E. (1972) *J. Biol. Chem.* **247**, 1338–1339

Hoffman, B. M., Roberts, J. E., Swansson, M., Speck, S. H. and Margoliash, E. (1980) *Proc. Natl Acad. Sci. U.S.A.* **77**, 1452–1456

Hopfield, J. J. (1974) *Proc. Natl Acad. Sci. U.S.A.* **71**, 3640–3644

Horie, S. and Morrison, M. (1963) *J. Biol. Chem.* **238**, 2859–2865

Höchli, L. and Hackenbrock, C. R. (1978) *Biochemistry* **17**, 3712–3719

Höjeberg, B. and Rydström, J. (1977) *Biochem. Biophys. Res. Commun.* **78**, 1183–1190

Hu, V., Chan, S. and Brown, G. (1977) *Proc. Natl Acad. Sci. U.S.A.* **74**, 3821–3825

Jacobs, E. E. and Sanadi, D. R. (1960) *J. Biol. Chem.* **235**, 531–534

Jacobs, E. E., Andrews, E. C., Cunningham, W. and Crane, F. L. (1966*a*) *Biochem. Biophys. Res. Commun.* **25**, 87–94

Jacobs, E. E., Kirkpatrick, F. H., Jr, Andrews, E. C., Cunningham, W. and Crane, F. L. (1966*b*) *Biochem. Biophys. Res. Commun.* **25**, 96–104

Jones, R. W., Lamont, A. and Garland, P. B. (1980) *Biochem. J.* **190**, 79–94

Jost, P. C., Griffiths, O. H., Capaldi, R. A. and Vanderkooi, G. (1973) *Proc. Natl Acad. Sci. U.S.A.* **70**, 480–484

Junge, W. and DeVault, D. (1975) *Biochim. Biophys. Acta* **408**, 200–214

Kagawa, Y. (1972) *Biochim. Biophys. Acta* **265**, 297–338

Karlsson, B. and Andréasson, L.-E. (1981) *Biochim. Biophys. Acta* **635**, 73–80

Kayalar, C. (1979) *J. Membr. Biol.* **45**, 37–42

Kayalar, C., Rosing, J. and Boyer, P. D. (1977) *J. Biol. Chem.* **252**, 2486–2491

Kawato, S., Sigel, E., Carafoli, E., Cherry, K. J. (1980) *J. Biol. Chem.* **255**, 5508–5510

Keilin, D. (1966) "The History of Cell Respiration and Cytochromes" (prepared by J. Keilin), Cambridge University Press, Cambridge

Keilin, D. and Hartree, E. F. (1939) *Proc. R. Soc. London B* **127**, 167–191

Keirns, J. J., Yang, C. S. and Gilmour, M. V. (1971) *Biochem. Biophys. Res. Commun.* **45**, 835–841

Kell, D. B. (1979) *Biochim. Biophys. Acta* **549**, 55–99.

Kessler, R. J., Blondin, G. A., Vande Zande, H., Haworth, R. A. and Green, D. E. (1977) *Proc. Natl Acad. Sci. U.S.A.* **74**, 3663–3666

Kilpatrick, L. and Erecińska, M. (1977) *Biochim. Biophys. Acta* **460**, 346–363

King, T. E., Mason, H. S. and Morrison, M. (eds) (1965) "Oxidases and Related Redox Systems", Vols 1 and 2, Wiley, New York

King, T. E., Orii, Y., Chance, B. and Okunuki, K. (eds) (1979) "Cytochrome Oxidase", Elsevier/North-Holland, Amsterdam.

Klingenberg, M. and Schollmeyer, P. (1961) *Biochem. Z.* **335**, 243–262

Komai, H. and Capaldi, R. A. (1973) *FEBS Lett.* **30**, 273–276

Kornblatt, J. A. and Williams, G. R. (1975) *Can. J. Biochem.* **53**, 467–471

Kornblatt, J. A., Kells, D. I. C. and Williams, G. R. (1975) *Can. J. Biochem.* **53**, 461–466

Kozlov, I. A. and Skulachev, V. P. (1977) *Biochim. Biophys. Acta* **463**, 29–89

Krab, K. (1977) Electron transfer and energy conservation at site 3 of oxidative phosphorylation. Ph.D. thesis, University of Amsterdam

Krab, K. and Slater, E. C. (1979) *Biochim. Biophys. Acta* **547**, 58–69

Krab, K. and Wikström, M. (1978) *Biochim. Biophys. Acta* **504**, 200–214

Krab, K. and Wikström, M. (1979) *Biochim. Biophys. Acta* **548**, 1–15

Kröger, A. (1978) in "Energy Conservation in Biological Membranes" (G. Schäfer and M. Klingenberg, eds), Springer-Verlag, Berlin, pp. 96–105

Kuboyama, M., Yong, F. C. and King, T. E. (1972) *J. Biol. Chem.* **247**, 6375–6383

Kunze, V. and Junge, W. (1977) *FEBS Lett.* **80**, 429–434

Lang, G., Lippard, S. J. and Rosén, S. (1974) *Biochim. Biophys. Acta* **336**, 6–14

Lanne, B. and Vänngård, T. (1978) *Biochim. Biophys. Acta* **501**, 449–457

Lanne, B., Malmström, B. G. and Vänngård, T. (1979) *Biochim. Biophys. Acta* **545**, 205–214

Lazdunski, M. (1972) *Eur. Top. Cell. Regul.* **6**, 267–310

Läuger, P. (1979) *Biochim. Biophys. Acta* **552**, 143–161

Lawford, H. G. (1978) *Can. J. Biochem.* **56**, 13–22

Lehninger, A. L., Ul Hassan, M. and Sudduth, H. C. (1954) *J. Biol. Chem.* **210**, 911–922

Leigh, J. S., Jr, Wilson, D. F., Owen, C. S. and King, T. E. (1974) *Arch. Biochem. Biophys.* **160**, 476–486

Lemberg, M. R. (1969) *Physiol. Revs* **49**, 48–121

Lemberg, R. and Gilmour, M. V. (1967) *Biochim. Biophys. Acta* **143**, 500–517

Levitzki, A. and Koshland, D. E., Jr (1976) *Curr. Top. Cell Regul.* **10**, 1–40

Lewis, A., Marcus, M. A., Ehrenberg, B. and Crespi, H. (1978) *Proc. Natl Acad. Sci. U.S.A.* **75**, 4642–4646

Lindsay, J. G. (1974) *Arch. Biochem. Biophys.* **163**, 705–715

Lindsay, J. G., Owen, C. S. and Wilson, D. F. (1975) *Arch. Biochem. Biophys.* **169**, 492–505

Longmuir, K. J., Capaldi, R. A. and Dahlquist, F. W. (1977) *Biochemistry* **16**, 5746–5755

Lorusso, M., Capuano, F., Boffoli, D., Stefanelli, R. and Papa, S. (1979) *Biochem. J.* **182**, 133–147

Love, B., Chan, S. H. P. and Stotz, E. (1970) *J. Biol. Chem.* **245**, 6664–6668

Ludwig, B. (1980) *Biochim. Biophys. Acta* **594**, 177–189

Ludwig, B. and Schatz, G. (1980) *Proc. Natl Acad. Sci. U.S.A.* **77**, 196–200

Ludwig, B., Downer, N. W. and Capaldi, R. A. (1979) *Biochemistry* **18**, 1401–1407

Machleidt, B. and Werner, S. (1979) *FEBS Lett,* **107**, 327–330
MacLennan, D. H. and Tzagoloff, A. (1965) *Biochim. Biophys Acta* **96**, 166–168
Makinen, M. W. (1979) in "Biochemical and Clinical Aspects of Oxygen" (W. S. Caughey, ed.), Academic Press, New York, pp. 143–155
Maley, G. F. and Lardy, H. A. (1954) *J. Biol. Chem.* **210**, 903–909
Malkin, R. and Malmström, B. G. (1970) *Adv. Enzymol.* **33**, 177–244
Malmström, B. G. (1973) *Quart. Rev. Biophys.* **6**, 389–431
Malmström, B. G. (1979) *Biochim. Biophys. Acta* **549**, 281–303
Malmström, B. G., Karlsson, B., Aasa, R., Andréasson, L.-E., Clore, G. M. and Vänngård, T. (1981) in "Interaction Between Iron and Proteins in Oxygen and Electron Transport" (C. Ho and W. C. Eaton, eds), Elsevier, New York, in press
Marres, C. A. M. and Slater, E. C. (1977) *Biochim. Biophys. Acta* **462**, 531–548
Marsh, D., Watts, A., Maschle, W. and Knowles, P. F. (1978) *Biochem. Biophys. Res. Commun.* **81**, 397–402
Mason, T. L. and Schatz, G. (1973) *J. Biol. Chem.* **248**, 1355–1360
Mason, T. L., Poyton, R. O., Wharton, D. C. and Schatz, G. (1973) *J. Biol. Chem.* **248**, 1345–1354
Merle, P. and Kadenbach, B. (1980) *Eur. J. Biochem.* **105**, 499–507
Minnaert, K. (1961) *Biochim. Biophys. Acta* **50**, 23–34
Minnaert, K. (1965) *Biochim. Biophys. Acta* **110**, 42–56
Mitchell, P. (1961) *Nature* **191**, 144–148
Mitchell, P. (1966) "Chemiosmotic Coupling in Oxidative and Photosynthetic Phosphorylation", Glynn Research Ltd, Bodmin, UK
Mitchell, P. (1968) "Chemiosmotic Coupling and Energy Transduction", Glynn Research Ltd, Bodmin, UK
Mitchell, P. (1977) *FEBS Lett.* **78**, 1–20
Mitchell, P. (1979) *Eur. J. Biochem.* **95**, 1–20
Mitchell, P. and Moyle, J. (1967) in "Biochemistry of Mitochondria" (E. C. Slater, Z. Kaniuga and L. Wojtczak, eds), Academic Press/PWN, New York and Warsaw, pp. 53–74
Mitchell, P. and Moyle, J. (1969) *Eur. J. Biochem.* **7**, 471–484
Mitchell, P. and Moyle, J. (1970) in "Electron Transport and Energy Conservation" (J. M. Tager *et al.*, ed), Adriatica Editrice, Bari, Italy, pp. 575–587
Mitchell, P. and Moyle, J. (1979) *Trans. Biochem. Soc.* **7**, 887–894
Mochan, I. and Nicholls, P. (1972) *Biochim. Biophys. Acta* **267**, 309–319
Monod, J., Wyman, J. and Changeux, J.-P. (1965) *J. Mol. Biol.* **12**, 88–118
Morrison, M. and Horie, S. (1964) *J. Biol. Chem.* **239**, 1432–1440
Morowitz, H. J. (1978) *Amer. J. Physiol.* **235**, R99–R114
Moss, T. H., Shapiro, E., King, T. E., Beinert, H. and Hartzell, C. R. (1978) *J. Biol. Chem.* **253**, 8072–8073
Muijsers, A. O., Slater, E. C. and Van Buuren, K. J. H. (1968) in "Structure and Function of Cytochromes" (K. Okunuki, *et al.*, eds), University of Tokyo Press, Tokyo, pp. 129–137
Muijsers, A. O., Tiesjema, R. H. and Van Gelder, B. F. (1971) *Biochim. Biophys. Acta* **234**, 481–492
Muijsers, A. O., Tiesjema, R. H., Henderson, R. W. and Van Gelder, B. F. (1972) *Biochim. Biophys. Acta* **267**, 216–221
Myer, Y. P. (1971) *J. Biol. Chem.* **246**, 1241–1248
Myer, Y. P. (1972) *Biochem. Biophys. Res. Commun.* **49**, 1194–1200

Nagle, J. F., Morowitz, H. J. (1978) *Proc. Natl Acad. Sci. U.S.A.* **75**, 298–302

Nagasawa, T., Nagasawa-Fujimori, H. and Heinrich, P. C. (1979) *Eur. J. Biochem.* **94**, 31–39

Nargang, F. E., Bertrand, H. and Werner, S. (1979) *Eur. J. Biochem.* **102**, 297–307

Nicholls, P. (1974*b*) in "Dynamics of Energy-transducing Membranes" (L. Ernster *et al.*, eds), Elsevier, Amsterdam, pp. 39–50

Nicholls, P. (1974*c*) *Biochim. Biophys. Acta* **346**, 261–310

Nicholls, P. (1975) *Biochim. Biophys. Acta* **396**, 24–35

Nicholls, P. (1976) *Biochim. Biophys. Acta* **430**, 13–29

Nicholls, P. (1978) *Biochem. J.* **175**, 1147–1150

Nicholls, P. (1979*a*) in "Biochemical and Clinical Aspects of Oxygen" (W. S. Caughey, ed.), Academic Press, New York, pp. 323–335

Nicholls, P. (1979*b*) *Biochem. J.* **183**, 519–529

Nicholls, P. and Chance, B. (1974) in "Molecular Mechanisms of Oxygen Activation" (D. Hayashi, ed.), Academic Press, New York, pp. 479–534

Nicholls, P. and Hildebrandt, V. (1978) *Biochem. J.* **173**, 65–72

Nicholls, P. and Kimelberg, H. K. (1972) in "Biochemistry and Biophysics of Mitochondrial Membranes" (G. F. Azzone *et al.*, eds), Academic Press, New York, pp. 17–32

Nicholls, P. and Petersen, L. C. (1974) *Biochim. Biophys. Acta* **357**, 462–467

Nicholls, P., Petersen, L. C., Miller, M. and Hansen, F. B. (1976) *Biochim. Biophys. Acta* **449**, 188–196

Ohnishi, T. (1975) *Eur. J. Biochem.* **64**, 91–103

Ohnishi, T., Blum, H., Leigh, J. S., Jr and Salerno, J. C. (1979) in "Membrane Bioenergetics" (C. P. Lee *et al.*, eds), Addison-Wesley, Reading, Mass., pp. 21–30

O'Keeffe, D. H., Barlow, C. H., Smythe, G. A., Fuchsman, W. H., Moss, T. H., Lilienthal, H. R. and Caughey, W. S. (1975) *Bio-inorg. Chem.* **5**, 125–147

Ondrias, M. R. and Babcock, G. T. (1980) *Biochem. Biophys. Res. Commun.* **93**, 29–35

Orii, Y. and King, T. E. (1972) *FEBS Lett.* **21**, 199–202

Orii, Y. and King, T. E. (1976) *J. Biol. Chem.* **251**, 7487–7493

Ozawa, T., Okumura, M. and Yagi, K. (1975) *Biochem. Biophys. Res. Commun.* **65**, 1102–1107

Ozawa, T., Tada, M. and Suzuki, H. (1979) in "Cytochrome Oxidase" (T. E. King, Y. Orii, B. Chance and K. Okunuki, eds), Elsevier/North-Holland, Amsterdam, pp. 39–52

Palmer, G., Babcock, G. T. and Vickery, L. E. (1976) *Proc. Natl Acad. Sci. U.S.A.* **73**, 2206–2210

Papa, S. (1976) *Biochim. Biophys. Acta* **456**, 39–84

Papa, S., Guerrieri, F., Lorusso, M., Izzo, G., Boffoli, D., Capuano, F., Capitanio, N. and Altamura, N. (1980) *Biochem. J.* **192**, 203–218

Phan, S. H. and Mahler, H. R. (1976*a*) *J. Biol. Chem.* **251**, 257–269

Phan, S. H. and Mahler, H. R. (1976*b*) *J. Biol. Chem.* **251**, 270–276

Peisach, J. (1978) in "Frontiers of Biological Energetics" (P. L. Dutton *et al.*, eds), Academic Press, New York, pp. 873–881

Penttilä, T. and Wikström, M. (1981) in "Vectorial Reactions in Electron and Ion Transport in Mitochondria and Bacteria" (F. Palmieri *et al.*, eds), Elsevier/North-Holland, Amsterdam, in press

Penttilä, T., Saraste, M. and Wikström, M. (1979) *FEBS Lett.* **101**, 295–300

Petersen, L. C. and Andréasson, L.-E. (1976) *FEBS Lett.* **66**, 52–57
Petty, R. H., Welch, B. R., Wilson, L. J., Bottomley, L. A. and Kadish, K. M. (1980) *J. Amer. Chem. Soc.* **102**, 611–620
Powers, L., Blumberg, W. E., Chance, B., Barlow, C. H., Leigh, J. S., Jr, Smith, J., Yonetani, T., Vik, S. and Peisach, J. (1979) *Biochim. Biophys. Acta* **546**, 520–538
Powers, L., Chance, B., Ching, Y. and Angiolillo, P. (1981) *Biophys. J.* **34**, 465–498
Poyton, R. O. and Schatz, G. (1975a) *J. Biol. Chem.* **250**, 752–761
Poyton, R. O. and Schatz, G. (1975b) *J. Biol. Chem.* **250**, 762–766
Poyton, R. O., McKemmie, E. and George-Nascimento, C. (1978) *J. Biol. Chem.* **253**, 6303–6306
Pozzan, T., Miconi, V., Di Virgilio, F. and Azzone, G. F. (1979) *J. Biol. Chem.* **254**, 10200–10205
Prochaska, L., Bisson, R. and Capaldi, R. A. (1980) *Biochemistry* **19**, 3174–3179
Prochaska, L. J., Steffens, G. C. M., Buse, G. M., Bisson, R. and Capaldi, R. A. (1981) *Biochim. Biophys. Acta* **637**, 360–373
Rascati, R. J. and Parsons, P. (1979a) *J. Biol. Chem.* **254**, 1586–1593
Rascati, R. J. and Parsons, P. (1979b) *J. Biol. Chem.* **254**, 1594–1599
Reed, C. A. and Landrum, J. T. (1979) *FEBS Lett.* **106**, 265–267
Reinhammar, B., Malkin, R., Jensen, P., Karlsson, B., Andréasson, L.-E., Aasa, R., Vänngård, T. and Malmström, B, G. (1980) *J. Biol. Chem.* **255**, 5000–5003
Rieder, R. and Bosshard, H. R. (1978) *J. Biol. Chem.* **253**, 2045–2053
Rieder, R. and Bosshard, H. R. (1980) in "1st European Bioenergetics Conference: Short Reports", Patron Editore, Bologna, Italy, pp. 87–88
Robertson, R. N. (1960) *Biol. Rev.* **35**, 231–264
Robinson, J. D. and Flashner, M. S. (1979) *Biochim. Biophys. Acta* **549**, 145–176
Robinson, N. C. and Capaldi, R. A. (1977) *Biochemistry* **16**, 375–381
Robinson, N. S., Strey, F. and Talbert, L. (1980) *Biochemistry* **19**, 3656–3661
Rodkey, F. L. and Ball, E. G. (1950) *J. Biol. Chem.* **182**, 17–28
Rosén, S. (1978a) *Biochim. Biophys. Acta* **503**, 389–397
Rosén, S. (1978b) *Biochim. Biophys. Acta* **523**, 314–320
Rosén, S., Brändén, R., Vänngård, T. and Malmström, B. G. (1977) *FEBS Lett.* **74**, 25–30
Rosevear, P., Van Aken, T., Baxter, J. and Ferguson-Miller, S. (1980) *Biochemistry* **19**, 4108–4115
Rubin, M. S. and Tzagoloff, A. (1973a) *J. Biol. Chem.* **248**, 4269–4274
Rubin, M. S. and Tzagoloff, A. (1973b) *J. Biol. Chem.* **248**, 4275–4279
Rydström, J. (1977) *Biochim. Biophys. Acta* **463**, 155–184
Rydström, J. (1979) *J. Biol. Chem.* **254**, 8611–8619
Saari, H., Penttilä, T. and Wikström, M. (1980) *J. Bioenerg. Biomembr.* **12**, 325–338
Sacher, R., Steffens, G. J. and Buse, G. (1979) *Hoppe-Seyler's Z. Physiol. Chem.* **360**, 1385–1392
Salmeen, I., Rimai, L. and Babcock, G. T. (1978a) *Biochemistry* **17**, 800–806
Salmeen, I., Rimai, L. and Babcock, G. T. (1978b) in "Frontiers of Biological Energetics" (P. L. Dutton *et al.*, eds) Academic Press, New York, pp. 905–911
Saraste, M., Penttilä, T., Coggins, J. R. and Wikström, M. (1980) *FEBS Lett.* **114**, 35–38
Saraste, M., Penttilä, T. and Wikström, M. (1981) *Eur. J. Biochem.* **115**, 261–268

Schatz, G. and Mason, T. L. (1974) *Annu. Rev. Biochem.* **43**, 51–87

Schlieper, P. and DeRobertis, E., (1977) *Arch. Biochem. Biophys.* **124**, 204–208

Schroedl, N. A. and Hartzell, C. R. (1977a) *Biochemistry* **16**, 1327–1333

Schroedl, N. A. and Hartzell, C. R. (1977b) *Biochemistry* **16**, 4961–4965

Schroedl, N. A. and Hartzell, C. R. (1977c) *Biochemistry* **16**, 4966–4971

Sebald, W., Machleidt, W. and Otto, J. (1973) *Eur. J. Biochem.* **38**, 311–324

Sebald, W., Machleidt, W. and Wachter, E. (1980) *Proc. Natl Acad. Sci. U.S.A.* **77**, 785–789

Seelig, A. and Seelig, J. (1978) *Hoppe Seyler's Z. Physiol. Chem.* **359**, 1747–1756

Seiter, C. H. A., Angelos, S. G., Jr and Perreault, R. A. (1978) in "Frontiers of Biological Energetics" (P. L. Dutton *et al.*, eds), Academic Press, New York, pp. 897–903

Seiter, C. H. A. and Angelos, S. G. (1980) *Proc. Natl Acad. Sci. U.S.A.* **77**, 1806–1808

Seki, S., Hayashi, G. and Oda, T. (1970) *Arch. Biochem. Biophys.* **138**, 110–121

Sekuzu, I., Takemori, S., Yonetani, T. and Okunuki, (1959) *J. Biochem.* (Tokyo) **46**, 43–49

Sevarino, K. A. and Poyton, R. O. (1980) *Proc. Natl Acad. Sci. U.S.A.* **77**, 142–146

Shapiro, A. L., Vinuela, E. and Maizel, J. V., Jr (1967) *Biochem. Biophys. Res. Commun.* **28**, 815–820

Sharrock, M. and Yonetani, T. (1977) *Biochim. Biophys. Acta* **462**, 718–730

Shaw, R. W., Hansen, R. E. and Beinert, H. (1978) *Biochim. Biophys. Acta* **504**, 187–199

Shaw, R. W., Hansen, R. E. and Beinert, H. (1979) *Biochim. Biophys. Acta* **548**, 386–396

Shaw, R. W., Rife, J. E., O'Leary, M. H. and Beinert, H. (1981) *J. Biol. Chem.* **256**, 1105–1107

Sigel, E. (1980) Functional studies on mitochondrial cytochrome *c* oxidase in natural and reconstituted membrane systems. Dissertation, Swiss Federal Institute of Technology, Zürich

Sigel, E. and Carafoli, E. (1978) *Eur. J. Biochem.* **89**, 119–123

Sigel, E. and Carafoli, E. (1979) *J. Biol. Chem.* **254**, 10572–10574

Sigel, E. and Carafoli, E. (1980) *Eur. J. Biochem.* **111**, 299–306

Slater, E. C., Van Gelder, B. F. and Minnaert, K. (1965) in "Oxidases and Related Redox Systems" (T. E. King, H. S. Mason and M. Morrison, eds), Wiley, New York, pp. 667–700

Smith, H. T., Standenmayer, N. and Millet, F. (1977) *Biochemistry* **16**, 4971–4974

Smith, L. and Conrad, H. E. (1956) *Arch. Biochem. Biophys.* **63**, 403–413

Smith, L., Davies, H. C., Reichlin, M. and Margoliash, E. (1973) *J. Biol. Chem.* **248**, 237–243

Stannard, J. N. and Horecker, B. L. (1948) *J. Biol. Chem.* **172**, 599–608

Steffens, G. J. and Buse, G. (1976) *Hoppe-Seyler's Z. Physiol. Chem.* **357**, 1125–1137

Steffens, G. J. and Buse, G. (1979) *Hoppe-Seyler's Z. Physiol. Chem.* **360**, 613–619

Steffens, G. C. M., Steffens, G. J. and Buse, G. (1979) *Hoppe-Seyler's Z. Physiol. Chem.* **360**, 1641–1650

Steffens, G. J., Buse, G., Steffens, G. C. M., Sacher, R. and Erdweg, M. (1981) in "Interaction Between Iron and Proteins in Oxygen and Electron Transport" (C. Ho and W. C. Eaton, eds), Elsevier, New York, in press

Steitz, T. A., Fletterick, R. J., Anderson, W. F. and Anderson, C. A. (1976) *J. Mol. Biol.* **104**, 197–222

Stevens, T. H., Brudwig, G. W., Bocian, D. F. and Chan, S. J. (1979*a*) *Proc. Natl Acad.Sci. U.S.A.* **76**, 3320–3324

Stevens, T. H., Bocian, D. F. and Chan, S. I. (1979*b*) *FEBS Lett.* **97**, 314–316

Stucki, J. W. (1978) in "Energy Conservation in Biological Membranes" (G. Schäfer and M. Klingenberg, eds), Springer-Verlag, Berlin, pp. 264–287

Sun, F. F., Prezbindowski, K. S., Crane, F. L. and Jacobs, E. E. (1968) *Biochim. Biophys. Acta* **153**, 804–818

Swank, R. T. and Munkres, K. D. (1971) *Anal. Biochem.* **39**, 462–477

Swanson, M. S., Quintanilha, A. T. and Thomas, D. D. (1980) *J. Biol. Chem.* **255**, 7494–7502

Tanaka, M., Haniu, M., Yasunobu, K. T., Yu, C. A., Yu, L., Wei, Y. H. and King, T. E. (1979) *J. Biol. Chem.* **254**, 3879–3885

Tanford, C. and Reynolds, J. A. (1976) *Biochim. Biophys. Acta* **457**, 133–170

Tanford, C., Nozaki, Y., Reynolds, J. A. and Makino, S. (1974) *Biochemistry* **13**, 2369–2376

Taylor, C. P. S. (1977) *Biochim. Biophys. Acta* **491**, 137–149

Thalenfeld, B. E. and Tzagoloff, A. (1980) *J. Biol. Chem.* **255**, 6173–6180

Thomson, A. J., Brittain, T., Greenwood, C. and Springall, J. (1976) *FEBS Lett.* **67**, 94–98

Tiesjema, R. H. and Van Gelder, B. F. (1974) *Biochim. Biophys. Acta.* **347**, 202–214

Tiesjema, R. H., Muijsers, A. O. and Van Gelder, B. F. (1972) *Biochim. Biophys. Acta* **256**, 32–42

Tiesjema, R. H., Muijsers, A. O. and Van Gelder, B. F. (1973) *Biochim. Biophys. Acta* **305**, 19–28

Tracy, R. P. and Chan, S. H. P. (1979) *Biochim. Biophys. Acta* **576**, 109–117

Tsudzuki, T. and Wilson, D. F. (1971) *Arch. Biochem. Biophys.* **145**, 149–154

Tweedle, M. F., Wilson, L. J., Garcia-Iniquez, L., Babcock, G. T. and Palmer, G. (1978) *J. Biol. Chem.* **253**, 8065–8071

Tzagoloff, A. and Wharton, D. C. (1965) *J. Biol. Chem.* **240**, 2628–2633

Urry, D. W., Wainio, W. W. and Grebner, D. (1967) *Biochem. Biophys. Res. Commun.* **27**, 625–631

Vail, W. J. and Riley, R. K. (1974) *FEBS Lett.* **40**, 269–273

Van Buuren, K. J. H. (1972) Binding of cyanide to cytochrome aa_3. Ph.D. thesis, University of Amsterdam

Van Buuren, K. J. H., Nicholls, P. and Van Gelder, B. F. (1972) *Biochim. Biophys. Acta* **256**, 258–276

Van Buuren, K. J. H. Van Gelder, B. F., Wilting, J. and Braams, R. (1974) *Biochim. Biophys. Acta* **333**, 421–429

Vanderkooi, G. (1974) *Biochim. Biophys. Acta* **344**, 307–345

Vanderkooi, J. and Erecińska, M. (1976) in "The Enzymes of Biological Membranes" Vol. 4 (A. Martonosi, ed.), Wiley, New York, pp. 43–86

Vanderkooi, G., Senior, A. E., Capaldi, R. A. and Hayashi, H. (1972) *Biochim. Biophys. Acta* **274**, 38–48

Vanderkooi, J. M., Landesberg, R., Hayden, G. W. and Owen, C. S. (1977) *Eur. J. Biochem.* **81**, 339–347

Van Gelder, B. F. (1966) *Biochim. Biophys. Acta* **118**, 36–46

Van Gelder, B. F. and Beinert, H. (1969) *Biochim. Biophys. Acta* **189**, 1–24

Van Gelder, B. F. and Muijsers, A. O. (1966) *Biochim. Biophys. Acta* **118**, 47–57

Van Gelder, B. F. and Slater, E. C. (1963) *Biochim. Biophys Acta* **73**, 663–665

Van Gelder, B. F., Van Rijn, J. L. M. L., Schilder, G. J. A. and Wilms, J. (1977) in "Structure and Function of Energy-transducing Membranes" (K. van Dam and B. F. van Gelder, eds), Elsevier/North-Holland, Amsterdam, pp. 61–68

Vänngård, T.(1972) in "Biological Applications of EPR" (H. M. Swartz *et al.*, eds) Wiley, New York, pp. 411–447

Vanneste, W. H. (1966) *Biochemistry* **5**, 838–848

Vanneste, W. H., Ysebaert-Vanneste, M. and Mason, H. S. (1974) *J. Biol. Chem.* **249**, 7390–7401

Van Verseveld, H. W., Krab, K. and Stouthamer, A. H. (1981) *Biochim. Biophys. Acta* **635**, 525–534

Vik, S. and Capaldi, R. A. (1977) *Biochemistry* **16**, 5755–5759

Vik, S. and Capaldi, R. A. (1980) Biochem. Biophys. Res. Commun. **94**, 348–354

Volpe, J. A., O'Toole, M. C. and Caughey, W. S. (1975) *Biochem. Biophys. Res. Commun.* **62**, 48–53

Wainio, W. W., Laskowska-Klita, T., Rosman, J. and Grebner, D. (1973) *Bioenergetics* **4**, 455–467

Wakabayashi, T., Senior, A. E., Hatase, O., Hayashi, H. and Green, D. E. (1972) *Bioenergetics* **3**, 339–344

Waltz, D. (1979) *Biochim. Biophys. Acta* **505**, 279–353

Warren, G. B., Honslay, M. D., Metcalfe, J. C. and Birdsall, N. J. M. (1975) *Nature* **255**, 684–687

Weber, K. and Osborn, M. (1969) *J. Biol. Chem.* **244**, 4406–4412

Weiss, H. and Sebald, W. (1978) *Methods Enzymol.* **53**, 66–73

Weiss, H. and Kolb, H. J. (1979) *Eur. J. Biochem.* **99**, 139–149

Weiss, H., Sebald, W. and Bucher, T. (1971) *Eur. J. Biochem.* **22**, 19–26

Weiss, H., Juchs, B. and Ziganke, B. (1978) *Methods Enzymol.* **53**, 66–73

Wever, R. and Van Gelder, B. F. (1974) *Biochim. Biophys. Acta* **368**, 311–317

Wever, R., Muijsers, A. O., Van Gelder, B. F., Bakker, E. P. and Van Buuren, K. J. H. (1973) *Biochim. Biophys. Acta* **325**, 1–7

Wever, R., Van Drooge, J. H., Van Ark, G. and Van Gelder, B. F. (1974) *Biochim. Biophys. Acta* **347**, 215–223

Wever, R., Van Gelder, B. F. and DerVartanian, D. V. (1975) *Biochim. Biophys. Acta* **387**, 189–193

Wever, R., Van Drooge, J. H., Muijsers, A. O., Bakker, E. P. and Van Gelder, B. F. (1977) *Eur. J. Biochem.* **73**, 149–154

Wharton, D. C. and Gibson, Q. H. (1968) *J. Biol. Chem.* **243**, 702–706

Wharton, D. C. and Tzagoloff, A. (1964) *J. Biol. Chem.* **239**, 2036–2041

Wikström, M. K. F. (1972) *Biochim. Biophys. Acta* **283**, 385–390.

Wikström, M. K. F. (1975) in "Electron Transfer Chains and Oxidative Phosphorylation" (E. Quagliariello *et al.*, eds), North-Holland/American Elsevier, Amsterdam, pp. 97–103

Wikström, M. K. F. (1977) *Nature* **266**, 271–273

Wikström, M. K. F. (1978) in "The Proton and Calcium Pumps" (G. F. Azzone *et al.*, eds), Elsevier/North-Holland, Amsterdam, pp. 215–226

Wikström, M. (1981*a*) *Curr. Top. Membr. Transport*, in press

Wikström, M. (1981*b*) in "Interaction Between Iron and Proteins in Oxygen and Electron Transport" (C. Ho and W. C. Eaton, eds), Elsevier, New York, in press

Wikström, M. (1981*c*) *Proc. Natl Acad. Sci. U.S.A.*, **78**, 4051–4054

Wikström, M. (1981*d*) in "Mitochondria and Microscomes" (C. P. Lee *et al.*, eds), Addison-Wesley, Reading, Mass., 249–269

Wikström, M. (1981*e*) in "The Proton Cycle" (V. P. Skulachev, and P. C. Hinkle eds), Addison-Wesley, Reading, Mass., in press

Wikström, M. and Krab, K. (1978) in "Energy Conservation in Biological Membranes" (G. Schäfer and M. Klingenberg, eds), Springer-Verlag, Berlin, pp. 128–139

Wikström, M. and Krab, K. (1979*a*) *Biochim. Biophys. Acta* **549**, 177–222

Wikström, M. and Krab, K. (1979*b*) *Biochem. Soc. Trans.* **7**, 880–887

Wikström, M. and Krab, K. (1979*c*) in "Cation Flux Across Biomembranes" (Y. Mukohata and L. Packer, eds), Academic Press, New York, pp. 321–329

Wikström, M. and Krab, K. (1980) *Curr. Top. Bioenerg.* **10**, 51–101

Wikström, M. K. F. and Saari, H. T. (1975) *Biochim. Biophys. Acta* **408**, 170–179

Wikström, M. K. F. and Saari, H. T. (1977) *Biochim. Biophys. Acta* **462**, 347–361

Wikström, M. K. F. and Saris, N.-E. L. (1970) in "Electron Transfer and Oxidative Phosphorylation" (J. M. Tager *et al.*, eds), Adriatica Editrice, Bari, pp. 77–88

Wikström, M. K. F., Harmon, H. J., Ingledew, W. J. and Chance, B. (1976) *FEBS Lett.* **65**, 259–277

Wikström, M., Saari, H., Penttilä, T. and Saraste, M. (1978) *FEBS Symp.* **45**, 85–94

Wikström, M., Krab, K. and Saraste, M. (1981) *Annu. Rev. Biochem.* **50**, 623–655

Williams, R. J. P. (1961) in "Haematin Enzymes" (J. E. Falk, R. Lemberg and R. K. Morton, eds), Pergamon Press, Oxford, pp. 41–53

Williams, G. R., Lemberg, R. and Cutler, M. E. (1968) *Can. J. Biochem.* **46**, 1371–1379

Wilson, D. F. (1967) *Biochim. Biophys. Acta* **131**, 431–440

Wilson, D. F. and Chance, B. (1966) *Biochem. Biophys. Res. Commun.* **23**, 751–756

Wilson, D. F. and Chance, B. (1967) *Biochim. Biophys. Acta* **131**, 421–430

Wilson, D. F. and Dutton, P. L. (1970) *Arch. Biochem. Biophys.* **136**, 583–584

Wilson D. F. and Leigh, J. S., Jr (1972) *Arch. Biochem. Biophys.* **150**, 154–163

Wilson, D. F. and Leigh, J. S. (1974) *Ann. N.Y. Acad. Sci.* **227**, 630–635

Wilson, D. F., Erecińska, M. and Brocklehurst, E. S. (1972*a*) *Arch. Biochem. Biophys.* **151**, 180–187

Wilson, D. F., Lindsay, J. G. and Brocklehurst, E. S. (1972*b*) *Biochim. Biophys. Acta* **256**, 277–286

Wilson, D. F., Dutton, P. L. and Wagner, M. (1973) *Curr. Top. Bioenerg.* **5**, 233–265

Wilson, D. F., Erecińska, M. and Owen, C. S. (1976) *Arch. Biochem. Biophys.* **175**, 160–173

Wilson, D. F., Owen, C. S. and Holian, A. (1977) *Arch. Biochem. Biophys.* **182**, 749–762

Wilson, M. T., Greenwood, C., Brunori, M. and Antonini, E. (1975) *Biochem. J.* **147**, 145–153

Wilson, M. T., Colosimo, A., Brunori, M. and Antonini, E. (1978) in "Frontiers of Biological Energetics (P. L. Dutton *et al.*, eds), Academic Press, New York, pp. 843–850

Wilson, M. T., Lalla-Maharajh, W., Darley-Usmar, V., Bonaventura, J., Bonaventura, C. and Brunori, M. (1980) *J. Biol. Chem.* **255**, 2722–2728

Winter, D. B., Bruyninckx, W. J., Foulke, F. G., Grinich, N. P. and Mason, H. S. (1980) *J. Biol. Chem.* **255**, 11408—11414

Wood, P. M. (1974) *FEBS Lett.* **44**, 22–24

Wrigglesworth, J. M., Baum, H. and Nicholls, P. (1973) *FEBS Lett.* **35**, 106–108

Yasunobo, K. T., Tanaka, M., Haniu, M., Samashima, M., Reimer, N. and Eto, T. (1979) in "Cytochrome Oxidase" (T. E. King *et al.*, eds), Elsevier, Amsterdam, pp. 91–101

Yamazaki, I., Yokota, K. and Tamura, M. (1966) in "Hemes and Hemoproteins" (B. Chance, R. W. Estabrook and T. Yonetani, eds), Academic Press, New York, pp. 319–325

Yatscoff, R. W., Freeman, K. B. and Vail, W. J. (1977) *FEBS Lett.* **81**, 7–9

Yonetani, T. (1960*a*) *J. Biol. Chem.* **235**, 845–852

Yonetani, T. (1960*b*) *J. Biol. Chem.* **235**, 3138–3143

Yonetani, T. (1961) *J. Biol. Chem.* **236**, 1680–1688

Yonetani, T. (1970) *Advan. Enzymol.* **33**, 309–335

Yonetani, T., Iizuka, T., Tamamoto, H. and Chance, B. (1973) in "Oxidases and Related Redox Systems II", Vol. 1 (T. E. King *et al.*,e ds), University Park Press, 401–405

Yoshikawa, S., Choc, M. G., O'Toole, M. C. and Caughey, W. S. (1977) *J. Biol. Chem.* **252**, 5498–5508

Young, L. J., Choc, M. G. and Caughey, W. S. (1979) in "Biochemical and Clinical Aspects of Oxygen" (W. S. Caughey, ed.), Academic Press, New York, pp. 355–360

Yu, C. A. and Yu, L. (1977) *Biochim. Biophys. Acta* **495**, 248–259

Yu, C. A., Yu, L. and King, T. E. (1975) *J. Biol. Chem.* **250**, 1383–1392

Yu, C. A., Yu, L. and King, T. E. (1977) *Biochem. Biophys. Res. Commun.* **74**, 670–676

Notes added in proof

1. It may be considered unlikely that employed redox mediators are able to equilibrate directly with the haem a_3/Cu_B centre, particularly in experiments with intact mitochondria. The recent finding of energy-dependent partial reversal of the oxygen reaction (Wikström, 1981c; Chapter 7, Section III.A) may therefore provide an explanation for the effect of ATP on the measured E_m values and extinctions of the aa_3 system in mitochondria (see Wilson $et\ al.$, 1972b; Wikström $et\ al.$, 1976). Although ATP decreases the apparent E_m values of both haems a and a_3, the effect is greater on the latter. Thus forms of a_3 other than reduced state are effectively stabilized in

the "high energy" situation (e.g. $Fe^{IV}=O^{2-}\ Cu_B^{II}-OH^-$ or $Fe^{III}-\overset{\displaystyle O}{\underset{\displaystyle O}{|}}-Cu_B^{II}$;

see Chapter 6 and pp. 155–157).

In a redox potential titration with ATP, therefore, the high-potential haem transition is mainly due to haem a (cf. the case with azide; Chapter 5), which explains its relatively high contribution to the 605 nm band in the energized state (cf. Wikström $et\ al.$, 1976). In this interpretation the apparent lack of effect of ATP on the E_m of Cu_B (Lindsay $et\ al.$, 1975) is still enigmatic, and this point requires further study.

2. The electron transfer activity of the isolated enzyme varies greatly with the experimental conditions (cf. Nicholls and Kimelberg, 1972). The maximum turnover number (moles of cytochrome c oxidized per second per mole of aa_3) at pH 7 and 25 °C may reach the order of 400 under optimal conditions. This is comparable to the maximal turnover number of the enzyme $in\ situ$ (see also Hartzell $et\ al.$, 1978; Rosevear $et\ al.$, 1980; Vik and Capaldi, 1980).

3. Green and Vande Zande (1981) incubated equimolar amounts of cytochrome oxidase with ferrocytochrome c in the presence of azide. They observed uptake of about one H^+ per ferrocytochrome c bound and interpreted the result in terms of protonation of ferrocytochrome c bound to the oxidase, in support of their thesis that oxidoreduction of bound cytochrome c is coupled to H^+ release and uptake. However, under their conditions nothing would prevent the transfer of electrons from ferrocyto-

190

chrome c to cytochrome a and Cu_A. Reduction of cytochrome a in the presence of azide is strongly pH-dependent (p. 109; Wilson *et al.*, 1976) and most electrons should reach haem a rather than Cu_A due to the high E_m of the former as compared to the latter in the presence of this inhibitor (p. 106). Uptake of H^+ is thus expected under the conditions employed by Green and Vande Zande and cannot be equated to H^+ uptake by cytochrome c without additional control experiments. The fact that the measured E_m of endogenous (bound) cytochrome c is independent of pH in the physiological range (Dutton *et al.*, 1970) was not taken into account by the former authors. This result shows that bound cytochrome c does not release or take up H^+ on oxidoreduction.

4. Fry and Green (1980) reported uptake of various cations into respiring cytochrome oxidase vesicles accompanied by anion uptake. These authors have placed considerable significance to this finding with respect to the mechanism of energy conservation (see Green and Vande Zande, 1981, and references therein). In contrast, we do not find significant translocation of either Ca^{2+} or K^+ by cytochrome oxidase vesicles (in the absence of added ionophores) at rates comparable with the rates of electron transfer by the enzyme. Since cytochrome oxidase activity generates an electrical potential difference across the liposomal membrane (Chapter 2) slow uptake of various ions may result at rates depending on the permeability of the membrane. However, we ascribe this to the generated membrane potential, but not to any specific property of cytochrome oxidase.

5. A recent X-ray absorption edge and EXAFS study of the isolated oxidase (Powers *et al.*, 1981) indicates a sulphur bridge between haem iron and copper of the a_3/Cu_B centre in the "resting" state, explaining the poor reactivity of this state with ligands (Chapter 4). However, due to the kinetic incompetence of the "resting" state with the catalytic mechanism (Chapter 6), the significance of the S bridge in catalysis remains uncertain, though it helps to explain findings that contradict the μ-oxo structure (see p. 59 and 62–63). At present, these EXAFS data do not necessitate any departure from our proposed mechanism of O_2 reduction (Fig. 6.6) other than regarding the structure of the "resting" state.

A remarkable conclusion from the EXAFS data is that the first shell structure of carbonmonoxy haem a_3 is identical to that of oxyhaemoglobin (except for the CO instead of O_2), This provides compelling support for the idea (p. 123) that dioxygen activation in the oxidase is due to fast electron transfer rather than to any structural peculiarity of haem iron or its proteinaceous vicinity.

Index

In part, this index serves also as a list of abbreviations, and as a complementary source of information.